THE WAR ON THE EPA

THE WAR ON THE EPA

America's Endangered Environmental Protections

William M. Alley
and
Rosemarie Alley

ROWMAN & LITTLEFIELD
Lanham • Boulder • New York • London

Published by Rowman & Littlefield
An imprint of The Rowman & Littlefield Publishing Group, Inc.
4501 Forbes Boulevard, Suite 200, Lanham, Maryland 20706
www.rowman.com

6 Tinworth Street, London SE11 5AL

British Library Cataloguing in Publication Information Available

Library of Congress Cataloging-in-Publication Data
Library of Congress Control Number: 2019953375

ISBN: 978-1-5381-3150-3 (cloth : alk. paper)
ISBN: 978-1-5381-3151-0 (electronic)

∞ ™ The paper used in this publication meets the minimum requirements
of American National Standard for Information Sciences—Permanence of
Paper for Printed Library Materials, ANSI/NISO Z39.48-1992.

This book is dedicated to the many current and past U.S. Environmental Protection Agency employees who are committed to protecting our environment.

About the cover: *Earthrise* is a photograph of the Earth and some of the Moon's surface that was taken on Christmas Eve, 1968, during the Apollo 8 mission, the first manned voyage to orbit the Moon. No one had ever seen an earthrise, and the emotional effect was a complete surprise to the Apollo 8 astronauts. Depicting the Earth as isolated and small against the vast blackness of space, it is one of the most iconic images of all time and inspirational for the environmental movement of the 1970s.

CONTENTS

ACKNOWLEDGMENTS

This book would never have been written without the help and insights of those who care about the U.S. Environmental Protection Agency and its role in people's lives. Suzanne Staszak-Silva at Rowman & Littlefield cheerfully guided us through each step in the publication process. Stacy Eisenstark, Lynn Goldman, Dave Jones, Alan Kolok, Jeff Kuhn, Paul Martin, John Mundell, Jovita Pajarito, Martin Lowenfish, Haley Waldkoetter, Barbara Wilson, John Wilson, and Mike Wireman all generously gave of their time for discussions and other input. The perspectives on the Environmental Protection Agency and any errors are ours alone. Finally, we're grateful to the Harvard Environmental and Energy Law Program for its Regulatory Rollback Tracker website, which documents the environmental regulatory rollbacks of the Trump administration. Circle of Blue and Aquafornia also provided useful daily coverage of important environmental news stories.

INTRODUCTION

Over the past couple of decades, Americans have been subjected to a systematic propaganda campaign to discredit science. Championed by ultra-conservatives, this campaign has come alarmingly close to accomplishing its goal by elevating people who don't have a clue what they're talking about to the same level as scientists who have worked long and hard to understand tough environmental problems. Having gained a foothold, propaganda is extremely difficult to reverse because of how it morphs with memory and learning. If you hear something enough, you're not only going to remember it, you're also more likely to believe it. Our survival is wired that way.

One of the most effective forms of propaganda is the *cherry-picking technique*. Richard Crossman, British deputy director of psychological warfare during World War II, explained why:

> It is a complete delusion to think of the brilliant propagandist as being a professional liar. The brilliant propagandist is the man who tells a selection of the truth in such a way that the recipient does not think he is receiving any propaganda. The art of propaganda is not telling lies, but rather selecting the truth you require and giving it mixed up with some truths the audience wants to hear.[1]

Today's propaganda campaign against science is focusing special attention on the U.S. Environmental Protection Agency (EPA), one of the most critical of federal agencies. What business, industry, and many Americans want to hear is that the EPA is hurting the economy, destroying jobs, and intruding into people's private lives. Lifted right out of the "cherry-picking" handbook, today's EPA is demonized for overregulation. There's some truth to these charges—the agency can be overly prescriptive—but the reality is much more complicated. Establishing any environmental regulation relies heavily on scientific findings and is a lengthy, challenging, and often futile undertaking.

Consider a few examples. Among the thousands of new chemicals, many of them highly toxic, none were added to drinking water regulations for over two decades. Agriculture is today's most pervasive source of water pollution yet is almost entirely unregulated. After nearly half a century, major battles are still being fought over which waterbodies are subject to regulation under the Clean Water Act. And then there's the EPA's role in regulating greenhouse gases under the Clean Air Act, which has set off the environmental version of the Civil War.

The Trump administration has become the most serious threat to the agency to date, but the war on the EPA can be traced back decades. The most far-reaching damage has come from undermining the foundation upon which the agency's legitimacy rests—its scientific capability and integrity.

This book takes the reader on a journey into some of today's most pressing environmental problems and how the EPA has become increasingly stymied in addressing them. We examine the science, politics, and human dimension of these issues. This is not an all-inclusive compendium of every problem facing the EPA or all that the agency does. It's also not a technical book about environmental policy or a history of the EPA—although we dip into these topics as needed. Rather, the purpose of this book is to explore the challenges of regulation and how the war on science is crippling the EPA's ability to regulate almost anything. What's at stake is our environmental protections.

1

EPA 101

Independent, honest science is the backbone of environmental regulation. It also threatens people who want to hide the truth.
—Christopher S. Zarba, former staff director of the U.S. Environmental Protection Agency Science Advisory Board[1]

The crisis that hit Toledo, Ohio, the night of August 1, 2014, was a lot like a major hurricane threatening an entire city and having to evacuate everyone. No exceptions. What happened in Toledo also involved everyone, almost half a million people. It was summertime, which means it was hot and humid in this midwestern city. Suddenly, around two in the morning, the mayor began issuing advisories through every available means of communication, and he wasn't mincing words—our drinking water has been poisoned so don't drink it, don't let your pets drink it, don't brush your teeth or prepare food with it, and don't boil it because that will only concentrate the toxins and make it worse.[2] Late-night revelers, tuned in to their favorite station on the way home, caught the breaking news.

Hours later, as Toledoans eased into their Saturday morning routines, the news was on the front page of the daily newspaper and pretty much everywhere on the internet. By now, the advisories had taken on darker implications. For example, as you were spitting out your morning coffee you also learned that you might have poisoned your three-year-old, who had woken up thirsty in the middle of the night. If symptoms appeared, the authorities got right to the point: seek medical attention. Immediate symptoms included diarrhea and vomiting, with liver and kidney damage also on the radar. Infants and children were the most vulnerable. And no one knew when the crisis would end. There was one bright spot, however—unlike people fleeing a hurricane, Toledoans knew they'd still have a home at the end of it.

The water ban lasted three days, which doesn't sound like such a big deal unless you're the one experiencing it. For one thing, it had no relation to the temporary inconvenience of waiting for the plumber to finish up and turn the water back on. And it couldn't be solved by dashing next door with a few empty containers when your water supply is suddenly, with absolutely no warning, cut off. Nor was the solution to hightail it to the nearest store, because every last bottled water in Toledo and neighboring communities had disappeared from the shelves. If you were the patient sort, you could get your name on a waiting list for the next bottled water delivery at Walmart or Walgreens or wherever, and then hunker down on the curb or on an overturned cart and wait it out. Governor John Kasich had called up the Ohio National Guard to haul water from all over the state to distribution centers at the city's high schools and fire departments, so that was another option. But there was still no getting around the patience requirement as life came to a virtual standstill and people stood in long lines, under a blazing August sun, for a few precious bottles of water.

The media reported pure pandemonium down at the water treatment plant, but to the extent that was true, it wasn't the running around and pulling your hair out variety. Chemists, lab technicians, and water honchos of all kinds were tweaking their water treatment cocktail, then sending off the latest sample (via a waiting plane) to the U.S. Environmental Protection Agency (EPA) lab in Cincinnati, and then resuming the waiting game until the results came back—hopefully somewhere in the normal range.

The problem was basically one of bad luck, due to a plantation-sized carpet of blue-green algae settling over the city's drinking water intake pipe out on Lake Erie. The algal bloom was generating a highly toxic poison called *microcystin*, which began showing up in the water treatment plant and then passing through the treatment barriers into the distribution system.

About five hundred years ago, the Swiss physician and chemist Paracelsus nailed the basic principle of toxicology when he said, "The dose makes the poison." One of the EPA's most important jobs is to figure out where this tipping point is for substances that pose a threat to human health and the environment, which allows them to set a standard for "safe" exposure. But when you consider all the thousands of natural and manmade substances that can be harmful, and what is involved in conducting toxicological and epidemiological studies that must consider different populations (infants, elderly, and healthy adults) and level of exposure (a single event, a few days, long-term)—all this is much easier said than done.

At the time of the Toledo crisis, the "safe" amount of microcystin in drinking water was based on the World Health Organization standard of *one part per billion*. Such a miniscule amount is hard to fathom, but it does give you an idea of how toxic this stuff is. During the water shutdown, they were getting measurements of 2.5 parts per billion.[3] A year after the crisis, the EPA announced a safe limit (health advisory) for short-term exposure of bottle-fed infants and preschool children of 0.3 parts per billion microcystin.

While water workers were experimenting with treatment options, there really was pandemonium at the mayor's and governor's offices. (You can use your imagination on this one.) Obviously, the most frequently asked question was some version of: When will the problem be fixed? Governor Kasich assured the city that the ban wouldn't be lifted until he was comfortable with his daughters and wife drinking the water, which was a little vague in terms of a timeline. Mayor Collins said he would not tell Toledoans to resume drinking water from their taps until he was convinced that it was safe for children. Again, vague. No doubt the second most frequently asked question was something like: What's causing the problem and what are we going to do about it so that it doesn't happen again?

The short answer to that perfectly logical question was some version of, "We're monitoring this very closely," which translates to there's not a darned thing we can do except cross our fingers and pray that those blue-green algae don't take up residence in our water neighborhood again. Fortunately, Mayor Collins (an independent), Governor Kasich (a Republican who is no fan of the EPA), and EPA officials worked closely throughout the ordeal.

Unbeknownst to most people, blue-green algae aren't algae—technically, they're *cyanobacteria*. It just happens that these bacteria are photosynthetic and look like algae. Cyanobacteria are generally considered to be the most successful group of microorganisms on earth. They are also the most genetically diverse, spanning the entire ballpark from good to bad. Cyanobacteria are credited with having accomplished the unique and impressive feat of oxygenating the planet. They are also such highly adaptable bacteria that they're found pretty much everywhere. These nimble little critters not only survive but *thrive* in saltwater or freshwater, in scalding hot springs, in salt formations, on rocks, and in soils. They've even been found in Antarctica. As long as their basic needs are met (that don't amount to much—a little phosphorus, a little nitrogen) they can grow, multiply, and bloom like an English spring. This is where their downside enters the picture. Some species of cyanobacteria produce toxins, the most common being *microcystin*. If that's not bad enough, they consume the oxygen in the water as they decompose, resulting in dead zones. There's some irony here, given that cyanobacteria were the original oxygen producers, resulting in the Great Oxygenation Event.

Exposure to cyanotoxins doesn't just come from drinking the water, but also from skin contact, breathing, or eating contaminated shellfish. Pets, livestock, and wildlife are also affected. Dogs don't hesitate to drink or swim in water that's on the green side, poisoning many every year. In Monterey Bay National Marine Sanctuary, twenty-one southern sea otters, a threatened species slowly returning from near extinction, died from eating shellfish laced with microcystin.[4]

The nutrients for algal blooms come from a variety of sources, but the primary culprit is fertilizers mixing with water and flowing off agricultural fields. What was once mostly a localized problem has now become a widespread and very serious national problem—somewhere in the same ballpark as the industrial dumping of yesteryear. In western Lake Erie, the problem is an unintended consequence of no-till farming, where fields are not plowed between plantings to prevent soil erosion. Consequently, much of the soluble phosphorus in the fertilizer remains at the surface, making it more likely to wash away into streams and eventually end up in the lake. Summers on Lake Erie are now referred to as "harmful algal bloom season." The problem will be made worse by climate change, as warmer water fosters the blooms and stronger storms flush more nutrients into lakes and rivers. Through an insidious feedback loop these blooms release methane and carbon dioxide, contributing to the greenhouse gas emissions causing climate change.

The 2014 drinking water crisis in Toledo was not the first time that algal toxins affected residents of Ohio or Lake Erie. In 2011, the largest algal bloom ever recorded on the lake could be seen from space—stretching 120 miles from Toledo to Cleveland. In 2013, an algal bloom temporarily shuttered a water treatment plant serving two thousand people in a town east of Toledo. In 2015, algal toxins affected drinking water and recreational activities in more than 650 miles of the Ohio River (two-thirds its length). Fortunately, there were no emergency shutdowns of water treatment plants.

Ohio isn't being singled out. Cyanotoxins have been implicated in human and animal illness, or death, in at least forty-three states.[5] A few make national news. In 2016, Florida Governor Rick Scott declared a state of emergency in Palm Beach and three neighboring counties because of a toxic algal bloom that spread to the beaches from Lake Okeechobee. The economic impact to locals was devastating. "This town is 100 percent driven by tourism but the tourism is empty," a surf-shop owner told CNN. "You go to the beach and it's the height of summer and we have empty beaches, empty restaurants, empty hotels. The smell is so horrible you have to wear a mask in the marina and the river."[6]

Lake Erie has come full circle. In the 1960s, the lake was declared "dead" and became a poster child for the 1972 Clean Water Act. By the 1980s, the lake had become an outstanding example of what humans can accomplish when they put their mind to it. One biologist described it as "the best example of ecosystem recovery in the world."[7] Today, it's once again a poster child for water pollution, this time because of harmful algal blooms. Voters in Toledo, Ohio, even passed the Lake Erie Bill of Rights in 2019. This groundbreaking law (soon challenged) allows the people of Toledo to act as legal guardians for Lake Erie, as if the citizens were the parents and the lake were their child. As such, citizens can sue polluters to pay for cleanup costs and prevention programs.[8]

The Toledo algal bloom not only illustrates how the EPA's efforts to protect our waterways and drinking water are far from over, it shows the difficulty of keeping ahead of emerging issues. Harmful algal blooms have been a growing concern for years, yet there were serious impediments affecting the response to the Toledo episode. Standardized methods for analysis of cyanotoxins, information on water treatment options, and guidance for communicating drinking water risks were all lacking. After the ordeal, the EPA addressed some of the unresolved issues. However, one prominent question remains: Should a drinking water standard for cyanotoxins be established, and if so, for which ones? A critical first step to answer this question involves nationwide monitoring of drinking water supplies that will conclude in 2020. In the next chapter, we'll look at why enacting drinking water regulations takes so long. For necessary context, let's first look briefly at the EPA's history and how the agency works.

A BRIEF HISTORY OF THE EPA

When President Nixon created the EPA in 1970, he didn't do it because he cared about the environment. He simply wanted to mollify the environmental "crazies" and divert national attention away from the Vietnam War and civil rights issues. In creating the EPA, it's been said that Nixon saw a mob coming, jumped in front of it, and called it a parade.[9]

And what a parade it was. During the 1970s, the U.S. Congress was the most progressive environmental legislative body in history. Within a decade, the country went from limited—and largely ineffective—environmental legislation at the state and federal level to major Congressional acts addressing air, water, and land. In 1970, Congress passed the Clean Air Act. In 1972, Congress passed the Clean Water Act and the Federal Insecticide, Fungicide, and Rodenticide Act. In 1974, Congress passed the Safe Drinking Water Act. In 1976, Congress passed the Resource Conservation and Recovery Act, which regulates hazardous wastes. That same year, Congress also passed the Toxic Substances Control Act. Finally, in 1980, in the midst of the Love Canal crisis, Congress passed Superfund. All these acts except Superfund were signed into law by Republican Presidents Nixon and Ford.

The tables turned on the EPA with the election of Ronald Reagan in 1980. Reagan had little interest in environmental issues, saying that if the EPA had its way, "you and I would live like rabbits."[10] He appointed Anne Gorsuch, a Colorado state legislator, as the EPA administrator. Gorsuch had led a successful battle to block her state's participation in the EPA's hazardous waste program and fought for less stringent auto emission rules. At the EPA, she became known as the "Ice Queen" for her frosty demeanor and hardline approach.[11]

A key tactic of Reagan's White House was to control regulatory agencies by putting the fox in charge of the hen house. Virtually all EPA appointees to key positions came from the very industries that the EPA was charged with overseeing. The other criterion for landing a plum job at the EPA was having virtually no experience (or interest) in the environment. An example is William Sullivan, a lawyer who often represented steel companies, who was put in charge of the EPA's national enforcement program. Sullivan later explained how he got the job: "I handled Reagan's stop in Youngstown as a candidate, and when they were recruiting, they asked for my resume. The EPA was the last job I wanted to go to, and enforcement was the last job I wanted at the agency."[12]

Given the continuing strong public support for environmental protections, the Reagan administration took a backdoor approach to neutralizing the agency. While insisting that she supported the EPA's programs, Gorsuch argued that she could carry out the agency's work better, and more efficiently, by streamlining resources. Along with massive budget reductions, Gorsuch slashed about a quarter of the workforce—helped along by a steady exodus of demoralized employees. Responsibility for many environmental programs was transferred to states, with virtually no accompanying federal oversight.

Three weeks after taking office, Gorsuch abolished the Office of Enforcement. Lawyers and staff who enforced air, water, hazardous waste, pesticide, and toxic chemicals laws were marginalized by distributing them across the agency. Soon after, a directive went out to the EPA regional offices that they should not send any cases to headquarters to be reviewed for possible action until they had explored "every opportunity for settlement." The note added, "Every case you do refer will be a black mark against you." The number of enforcement actions forwarded by the regional offices plummeted by 79 percent the following year.[13]

A pervasive mistrust of employees resulted in enormous morale problems. Gorsuch and her appointees gave all indications of believing what industry lobbyists told them over what highly qualified and committed EPA staff was saying. There were also the hit lists. An organizational chart of top career officials was created with a colored dot next to each name—blue for those deemed loyal to the administration, red for acceptable performers to be transferred to less responsible positions, and brown for those to be eased out. Employees referred to the chart as the Crayola code.[14] A former headquarters enforcement manager summed up the working conditions: "You were trying to survive, trying to continue to do your job, while most of your days were spent worrying about whether you would actually have a job, in some cases, or whom you would be working for and whether that person would be a rational human being."[15]

After about a year on the job, Gorsuch began to get into hot water. An investigation revealed that she had privately assured a small refiner that it wouldn't be penalized if it violated gasoline lead regulations.[16] She also kindled public outrage when she lifted a ban against dumping drums of liquid hazardous wastes in landfills. Many of Gorsuch's public relations problems stemmed from charges that the EPA was delaying Superfund cleanups for political reasons and making "sweetheart" deals with polluting companies to reduce their liabilities. In the fall of 1982, acting under orders from Reagan and the Justice Department, Gorsuch refused to provide Superfund documents that were subpoenaed by a congressional subcommittee. In response, the House of Representatives voted to hold Gorsuch in contempt—a first for a Cabinet-level officer.[17] Gorsuch resigned a few months later, after fewer than two years in office.

To address the contentious issues surrounding the agency, Reagan brought back William Ruckelshaus, the EPA's first administrator. Ruckelshaus was followed by Lee Thomas and then William Reilly under George H. W. Bush. All three administrators sought to bring a more scientific, risk-based approach to guide the agency's priorities. Ecological issues and pollution prevention also gained greater attention, as did international problems such as acid rain and the ozone hole.

During the 1980s, Congress renewed and strengthened every major environmental statute that came up for renewal.[18] These included the Resource Conservation and Recovery Act in 1984, the Safe Drinking Water Act and Superfund in 1986, and the Clean Water Act in 1987. The Clean Air Act underwent major revision and enhancement in 1990. And then, in 1994, the country's center of gravity changed when Newt Gingrich led the Republican takeover of the House for the first time in forty years—ushering in the end of a working bipartisan Congress. The war on the EPA and its science has intensified since that time.

The Clinton administration's most significant environmental achievement was standing firm against Republican attempts to roll back environmental laws and regulations. The only major EPA legislation under Clinton were revisions to pesticides and drinking water laws. Clinton's limited environmental record contributed to activist Ralph Nader's run for president in 2000 as a Green Party candidate, possibly serving as a spoiler for Al Gore in Florida and leading to George W. Bush's win. Immediately before leaving office, Clinton signed a whirlwind of environmental regulations. Bush promptly suspended Clinton's "midnight regulations," ultimately approving some of them (e.g., a revised drinking water standard for arsenic and limits on diesel emissions), but shelving or reversing others (e.g., cleanup requirements for mining on public lands). [19]

Enforcement activities and regulations dropped off significantly under Bush. As during the Gorsuch years, major environmental policy decisions were driven by political appointees who viewed their mission as the single-minded advancement of the president's policy agenda. As such, they ignored or downplayed the scientific advice and analysis of career employees. A signature controversy under Bush was accusations of censoring government scientists and altering their reports when these threatened the administration's lax environmental agenda in areas such as climate change and the listing of endangered and threatened species. [20]

A whole new level of antagonism emerged during the Obama presidency, with its environmental pushes in climate change, clean air and water, and renewable energy. Republicans in Congress fought back with every weapon at their disposal to delay, limit, or prevent EPA actions, declaring that the Obama administration was waging "an all-out assault on the American economy." Oklahoma Senator James Inhofe, Republican leader of the powerful Senate Energy and Natural Resources Committee, compared the EPA to Nazi Germany's Gestapo. In 2014, incoming Senate Majority Leader Mitch McConnell's (R-KY) number one priority was "to try to do whatever I can to get the EPA reined in." [21]

With the election of Donald Trump, anti-EPA fervor kicked into high gear. During his campaign, Trump repeatedly promised that he would get rid of the EPA. "We're going to have little tidbits left but we're going to get most of it out. What the EPA does is a disgrace," went one-such comment. When asked who will protect our environment, he quipped, "We'll be fine with the environment."[22]

By appointing Scott Pruitt as EPA administrator, Trump lost no time in making good on his promise. Even among Trump's most controversial appointments, Pruitt stood out as overtly hostile to the agency he was chosen to lead. During his six-year tenure as Oklahoma attorney general, Pruitt had led or took part in fourteen lawsuits against the EPA, particularly opposing regulations affecting his oil and gas donors.[23]

Pruitt operated out of the Gorsuch-era playbook: large proposed budget cuts under the guise of efficiency, key appointees coming from the business and industrial sectors that the EPA is charged with regulating, treating career EPA employees with disdain and ignoring their advice, undermining the balance between the role of states and the federal government, and making enforcement of environmental regulations a low priority. The idea was to roll back existing regulations and reduce the agency's ability so much that it wouldn't be able to recover even when the political winds change. He did this by relying largely on political appointees, former lobbyists, and industry officials to shape his policies.[24]

When Pruitt was first selected as the EPA administrator, the *New York Times* described him as "carefully plotting out a course to go after the EPA with a scalpel rather than a meat cleaver."[25] As soon became obvious, Pruitt viewed his job at the EPA as a launching platform to promote himself—possibly even making a run at the presidency in 2024.[26] Eager to make a big splash, Pruitt rushed through changes without building a legitimate case for many of his actions. As a result, many of his efforts to delay or roll back Obama-era regulations were struck down by the courts. Nonetheless, Pruitt did substantial damage to America's environmental protections, setting the stage for his successors to continue where he left off. He delayed rules that had not yet taken effect and set in motion rollbacks in regulations across the board. He demoralized the agency and caused many dedicated and highly qualified employees to leave the agency. He was a primary advocate for withdrawing from the

Paris climate agreement. He also inflicted substantial damage through omission.

Scott Pruitt brought new meaning to the abuse of high office: flying first-class and indulging in luxury hotel accommodations, enlisting staff to help his wife seek a fast-food franchise and later a high-paying job, and enjoying a fifty dollar-a-night sweetheart deal in a luxury Capitol Hill condo, co-owned by the wife of a lobbyist with interests in EPA decisions.[27] By spring 2018, Pruitt's belief that the rules didn't apply had caught up with him. With more than a dozen federal inquiries underway into his spending and ethical behavior, he resigned in September. Andrew Wheeler, a former coal lobbyist, assumed the reins as acting administrator. Much less flamboyant and politically ambitious than Pruitt, Wheeler continued the regulatory rollbacks. "I don't think the overall agenda is going to change that much," he told the *Washington Post*, "because we're implementing what the president has laid out for the agency."[28] Wheeler's subsequent actions would prove his point.

The Trump administration carried out an unprecedented effort to undermine the way in which science is used by government agencies. As an example, one of Pruitt's first acts was to put on hold a proposed EPA rule to ban the pesticide chlorpyrifos, a highly toxic chemical that causes brain damage and other neurological harm to children. After many years of research, scientists in the EPA's Office of Pesticide Programs were not consulted on the decision.[29] The Trump administration also targeted the EPA's scientific advisory committees. Under the guise of avoiding conflicts of interest, scientists who received EPA research grants were barred from serving on its scientific advisory committees—when, in fact, the scientific advisory boards don't decide on individual grants. At the same time, there was no such exclusion for experts working for industry, even if their firm is regulated by the EPA—a glaring conflict of interest.[30] Dozens of scientists on these committees were replaced with industry-friendly members. Even then, the committees were often bypassed when making key decisions.

It can be reassuring to know that we've been here before during the Gorsuch years and somehow survived, but the key difference is today's lack of bipartisan support for environmental regulations. A scorecard by the League of Conservation Voters that rates members of Congress on how they voted on environmental issues says it all. In 2017, Republican senators and House GOP members had average scores of 1 percent and 5 percent, respectively. Forty-six Republican senators and 124 House Republicans scored zero. In contrast, Democratic senators and representatives each averaged over 90 percent. Compounding the problem, with the exception of climate change, there is only lukewarm public concern about our endangered environmental protections.[31]

As the EPA nears the half-century mark, there's a reason why the American public is largely apathetic and silent. Overall, the environment looks like it's doing just fine. The highly visible problems, such as the crippling smog and filthy rivers of yesteryear, are now just a memory—if even that. Today's most serious environmental issues are largely invisible to the naked eye and more global in extent.

HOW THE EPA WORKS

As a large federal agency, the EPA has a unique organizational structure that allows it to work in partnership with the states. The EPA's first administrator, William Ruckelshaus, established the agency's headquarters in Washington, DC, as a practical necessity, and then created ten regional offices around the country—where most of the action takes place. As a result, the EPA is one of the most decentralized agencies in the federal government. Rule-making and policy development fall under the aegis of headquarters staff, but the regional offices, working in conjunction with the states (and Indian tribes), put policy to practice. It's a balancing game requiring both the carrot and, when all else fails, the stick. States are essential partners in environmental protection, but they also have competing interests—notably enticing industry to their state and keeping it there. A key challenge for the EPA has been how to delegate responsibility for environmental programs to the states, while remaining the ever-present "gorilla in the closet" (a term coined by Ruckelshaus) that assumes control if state authorities succumb to special interests or otherwise fail to do their job. Lead contamination in Flint, Michigan, the most high-profile environmental crisis in recent times, is a prime example of the EPA's regional office failing to take charge or bring out the headquarters' gorilla.

The EPA administrator and other top officials are political appointees who support the president's agenda. In contrast, the long-term career staff is focused on, and dedicated to, environmental protection. Outside scientists also play an important role. For example, the Science Advisory Board independently reviews the science behind some of the EPA's most consequential decisions and policies. The Clean Air Scientific Advisory Committee has a key role in determining air quality standards. Environmental and industry watchdog groups also keep a close eye on the EPA. The interplay among these groups can result in a challenging tension between politics and science.

At both the regional and headquarters levels, the EPA's organizational structure consists of separate program offices to address air pollution, surface water quality, groundwater and drinking water, pesticides, hazardous waste, and so forth. These offices operate largely independently, despite (as many of our examples demonstrate) the obvious need to work together to achieve pollution control.

The American paradigm of the EPA is reflected in the 1984 movie *Ghostbusters*. Walter Peck, the main antagonist, is a humorless, strictly by-the-book EPA inspector who shuts down the "unlicensed" storage system housing the Ghostbuster's mischievous spirits. The predictable massive explosion unleashes epic-scale consequences on New York City. Fortunately, in the real world, this is not the way the EPA operates.

The EPA's regulations are broadly defined by Congressional legislation, thereby giving the agency considerable latitude in establishing specific rules and how to go about enforcing them. Enforcement work is part sheriff with a badge and part diplomat at the table.

Limited staff and funding necessitate that EPA managers carefully choose their battles. Some violations can be quickly resolved, but complex cases often involve high stakes and hard choices. The EPA's enforcement staff face sensitive decisions involving acceptable pollution control measures, how much time to allow a violator to come into compliance, and the size of penalties. The last resort is always whether to refer the matter to the Justice Department for civil or criminal action. As an EPA regional supervisor put it, "You know, we really have to be very reasonable when we're in the enforcement business. The problem is that a lot of times it's just damned difficult figuring out what being reasonable *means*."[32]

John Quarles, deputy administrator under Ruckelshaus, explained the reality of taking intransigent violators to court: "Nearly everyone exposed for the first time to the realities of court action is astonished at the length of court dockets, the complexity of pretrial procedures, the opportunities for delay by opposing counsel, the time and effort required for preparing cases for trial, and numerous other difficulties in pushing a trial through to adjudication."[33] And then there are the appeals.

The EPA's first major test case set the tone for the agency's enforcement approach. In 1971, the EPA won a court case against Armco Steel for polluting the Houston Ship Channel, which had become a toxic waste sewer for Armco and other industries. Armco had been discharging over 975 pounds of cyanide, more than 380 pounds of phenols, and between six thousand and twelve thousand pounds of ammonia each day into the channel. The judge came down hard and prohibited *any* further discharge, effective *immediately*. Armco's chief executive officer wrote to Nixon, reminding the president that he had personally assured industry that they would not be a whipping boy in solving environmental problems. He added that the court order had eliminated about three hundred jobs in a stroke of the pen. The company then hunkered down and flatly refused to negotiate with the EPA. Newspapers got hold of the story and exposed Armco's campaign contributions to Nixon. The resulting public outcry forced the company to come to the table. How it played out was humane and reasonable. The EPA allowed the plant

to temporarily resume operations while it installed treatment facilities. (It also came out that those three hundred employees thrown
out of work by the EPA's meddling had, in fact, been laid off
several weeks before the judge's decision.)[34]

The Armco case set the stage for the EPA's approach in dealing
with many pollution cases—in short, aggressively pursue egregious
polluters and then negotiate a reasonable timeframe for compliance
that takes into consideration cost and impact on jobs. The EPA's
enforcement actions are often supplemented, or forced, by environmental groups acting through citizen lawsuits. Simultaneously,
many environmental policy professionals recommend greater adoption of more cooperative approaches and market incentives.

From the agency's first actions, challenging and legitimate questions arose about which takes priority, the environment or jobs.
Ruckelshaus was focused on this dilemma when he observed: "Public opinion remains *absolutely essential* for anything to be done on
behalf of the environment. Absent that, nothing will happen, because the forces of the economy and the impact on people's livelihood are so much more automatic and endemic." William Ruckelshaus held many prominent positions throughout his career, but he
ranked his tenure at the EPA above all others. "At EPA," he reflected, "you work for a cause that is beyond self-interest and larger
than the goals people normally pursue. You're not there for the
money."[35]

The EPA's organizational structure and initial policy decisions
serve as a starting point to understanding how the agency is addressing some of today's most serious environmental issues in the areas
of drinking water (chapters 2 and 3), water pollution (chapters 4 and
5), air pollution and climate change (chapters 6–8), and toxic chemicals and hazardous waste (chapters 9–12). The stories in each chapter illustrate the political, scientific, and regulatory challenges facing the agency.

Virtually everything that the EPA has accomplished has come
out of the crucible of intense controversy, with significant economic, health, and social consequences at stake. The agency almost
invariably finds itself entangled in major and long drawn out court

battles. In the scheme of things, this large federal agency charged with protecting our environment is, in reality, a David taking on the Goliath of big business. Even in the best of times, it's remarkable that anything gets done.

Part I

Drinking Water

2

TAKE IT FROM THE TAP

The most important healthcare provider in your community is the person who looks after your water.
—Bernadette Conant, chief executive officer, Canadian Water Network

In 1997, California water regulators got a major jolt. For decades, they had known that Aerojet, Lockheed Martin, and other defense contractors had dumped millions of gallons of perchlorate waste into unlined pits that had worked its way into groundwater. This had been a localized problem. With the availability of more sensitive laboratory techniques, the chemical was now showing up in drinking water far from these sources of contamination. The problem kicked into high gear when the Metropolitan Water District of Southern California started finding perchlorate in their drinking water plants. [1]

The Metropolitan Water District is the nation's largest supplier of drinking water, providing water to heavily populated portions of southern California. The agency employs almost two thousand people to keep this water megalith operating, including a battalion of scientists and technicians who perform hundreds of thousands of tests on water samples every year. The Metropolitan Water District of Southern California is so huge, with so much at stake, that it's known far and wide as simply MWD.

For MWD, the problem was two-fold. First, the health implications were alarming. Perchlorate interferes with the uptake of iodine by the thyroid. Reduced iodine uptake can lead to inadequate levels of thyroid hormone that helps regulate the body's metabolism and controls development of the central nervous system in fetuses and infants. Environmental health scientists generally agree that the risk of developing health problems from perchlorate contamination is low in healthy adults, but a prolonged decrease of thyroid hormone can have serious consequences in sensitive populations—pregnant women, fetuses, newborns, and people with thyroid disorders. Of most concern is that perchlorate in pregnant woman with low iodide levels can disrupt brain development in fetuses and infants. About one-third of U.S. women have these lower iodine levels.[2] The risk of perchlorate exposure to fetuses is greatest in the first trimester because brain development starts very early and is fully dependent on maternal hormone production. The developmental harm appears to be irreversible.[3]

MWD's second problem was that perchlorate was showing up in their plants largely supplied by the Colorado River—over two hundred miles to the east. A team of technicians was dispatched to Lake Havasu, the reservoir on the Colorado River that supplies much of southern California's drinking water. When they found perchlorate in the samples collected from the reservoir, this was no longer solely California's problem. Phoenix, Tucson, and other cities in Arizona are also heavily reliant on water from Lake Havasu.[4]

Additional sampling to locate the source found perchlorate in Lake Mead, 150 miles upriver from MWD's intake pipes. From there, officials tracked the contaminant to Las Vegas Wash, a tributary to Lake Mead. Further sampling discovered high concentrations of perchlorate in a seep that forms a small stream flowing into the wash. Like pulling on a string, they just followed the trail. The highest concentrations of perchlorate were found in groundwater three miles from the seep beneath a perchlorate plant operated by Kerr-McGee.[5]

The scope of the problem was staggering. A three-mile long plume was discharging about nine hundred pounds of perchlorate *every day* into Las Vegas Wash, with most of it ending up in Lake Mead.[6] This single source was contaminating the drinking water supply of fifteen million to twenty million people in Arizona, southern California, and southern Nevada. And it didn't end there. The winter lettuce crop for America's dinner tables, irrigated with Colorado River water, also contained perchlorate.[7]

This was not the first time that Kerr-McGee had a public relations problem. The movie *Silkwood* depicts one of the company's employees, Karen Silkwood, who died under mysterious circumstances while gathering evidence to implicate Kerr-McGee in exposing its workers to plutonium. The company also had a legacy of toxic contamination at chemical sites across the country. Rather than pay for the clean-up costs of these contaminated sites, Kerr-McGee spun off these assets into a separate company without revealing the full scope of the problems to investors. The Justice Department brought fraud charges against Kerr-McGee's successor company, Anadarko, to pay for the cleanups. In 2014, Anadarko settled for over five billion dollars, including $1.1 billion to a state of Nevada trust fund for cleaning up the perchlorate site. At the time, this was the largest recovery for the cleanup of environmental contamination in history.[8]

All told, the clean-up effort has been a success. Perchlorate entering Lake Mead has been reduced by 95 percent, with more than five thousand tons removed from the environment.[9] But the chemical continues to be found elsewhere around the country, often associated with military bases. Perchlorate makes solid rocket fuel burn (as well as flares and fireworks). It was an essential ingredient in building bigger and more powerful rockets during the Cold War and NASA's space shuttle. It remains critical to national defense, including today's Tomahawk and Minuteman missiles. The Navy uses it in its underwater munitions. There's a lot at stake here for the Department of Defense and NASA.[10]

Shortly after perchlorate was discovered in Lake Mead, the Department of Defense began testing for perchlorate in water and soils around their bases. From 1997 to 2009, the Department of Defense reported finding perchlorate at 284 (nearly 70 percent) of its installations sampled. Perchlorate also has been found at NASA sites, including the Jet Propulsion Laboratory in Pasadena, where local drinking water sources were contaminated. [11]

The U.S. Environmental Protection Agency (EPA) got involved in the Lake Mead issue in full knowledge that they were up against NASA and the world's most powerful war machine, the U.S. Department of Defense. The main battle line formed around what level of perchlorate can be considered safe. In 1998, the EPA set four to eighteen parts per billion as an "interim" range for perchlorate exposure while it completed a risk assessment. Four years later, the agency recommended a health-protective standard of one part per billion. With perchlorate showing up in military bases all over the country and billions of dollars of clean-up costs at stake, the Pentagon fought back, arguing for a standard of two hundred parts per billion. [12]

The Bush White House turned to the National Academy of Sciences to study the problem. The Academy report came out in 2005. Results of nationwide sampling also found less than one percent of public water systems exceeded fifteen parts per billion. The EPA relied on the Academy study to establish a goal of 24.5 parts per billion for cleanup at Superfund sites. A drinking water standard remained elusive, and states began to adopt their own standards. [13]

In October 2008, against the objections of its own scientists, the EPA administrator under the Bush administration opted *not* to regulate perchlorate, citing the need for more research. What followed was almost unheard of when an EPA advisory board on children's health issues posted a letter of protest on the agency's website. "This decision," the letter stated, "does not recognize the science which supports the exquisite sensitivity of the developing brain to even small drops in thyroid hormone levels" that could be caused by perchlorate. At risk, the scientists claimed, were "millions of pregnant women and their fetuses, and lactating women and infants across the country." The EPA's Science Advisory Board also weighed in, contending that the agency had acted hastily in giving the chemical a pass.[14]

In January 2009, during its last days in office, the Bush administration announced an interim health advisory for perchlorate of fifteen parts per billion. This value accounted for food as an additional source of perchlorate. A study by the U.S. Food and Drug Administration had found low levels of perchlorate in 74 percent of 285 food items tested. Certain foods, such as tomatoes and spinach, had higher levels than others.[15] The outgoing Bush administration also announced that it was asking the National Academy of Sciences to review the interim health advisory level—yet another in the long series of delays.

Senator Barbara Boxer, a firebrand liberal Democrat from California, was chair of the Environment and Public Works Committee, and she had had enough. For seven years, Boxer had been actively pushing for setting a perchlorate standard, so it could be regulated—enough of this soft-pedaling health advisory nonsense. "This is a widespread contamination problem, and to see the Bush EPA just walk away is shocking," she protested.[16] Less than a week after the EPA's announcement of yet another National Academy review, Boxer had a chance to do something about it. Lisa Jackson's confirmation hearing as Obama's first EPA administrator was coming up.

Jackson, as the former head of the New Jersey Department of Environmental Protection, was no stranger to the perchlorate controversy. In 2005, a panel of scientists, environmental activists, and industry leaders had urged New Jersey to regulate the chemical. Three years later, Jackson's department hadn't even completed a draft of the rule.[17]

Senator Boxer saved her ammunition for what she called a "lightning round" during the final minutes of Jackson's hearing. Boxer told Jackson that she wanted simple yes or no answers to a few questions. The first question was about perchlorate. After noting that California had 290 drinking water sources with at least four parts per billion of perchlorate, Boxer told Jackson:

Yet, EPA recently refused to regulate perchlorate. We had quite a to-do over here in that hearing. And they won't regulate it in drinking water, and they sent the issue back to the National Academy of Sciences. Now, again, delay, delay, delay. We have had years of it, and we need action! Do you commit to us to immediately review this failure to establish a drinking water standard for perchlorate and act to address the threat to pregnant women and children caused by this dangerous toxin?[18]

In accordance with Boxer's ground rules, Lisa Jackson simply replied, "Yes, Madam Chair."

In 2011, the Obama administration reversed the Bush administration's decision not to regulate perchlorate and committed to proposing a drinking water standard within two years.[19] The deadline came and went. The key issue, and continuing challenge, was how to determine the relationship between perchlorate exposure in drinking water and brain development in fetuses and infants. In 2013, the EPA Science Advisory Board recommended that the agency use a non-traditional approach known as physiologically based pharmacokinetic modeling. The EPA undertook development of this approach in collaboration with the U.S. Food and Drug Administration.

In 2016, with a perchlorate standard still nowhere in sight, the Natural Resources Defense Council took the EPA to court. Under a Consent Decree, the EPA committed to propose a standard by October 2018, with a six-month extension later granted. At long last, in May 2019, the EPA proposed the first new drinking water standard for a chemical in two decades—fifty-six parts per billion for perchlorate. The proposed standard was several times higher than the earlier health advisory of fifteen parts per billion and the science behind any standard continued to be controversial. The EPA also requested feedback on whether a drinking water standard was even needed. A final standard is due by June 2020.

Each year, every residence in the country that gets its water from a community water system receives a Consumer Confidence Report, which provides customers with information about their water sources, the contaminants in their water, and the health effects of these contaminants. This annual report was the brainchild of Congress when it was debating the 1996 Safe Drinking Water Act Amendments. It seemed a great idea at the time—a nice, simple, straightforward rundown on how your local water provider is doing in keeping your water safe to drink.

Water utility employees whose job it is to compile all this painstaking information may be passionate about this task, or they might just go through the motions with the same enthusiasm as filling out your tax return. But there's a crucial difference. Tax returns get read, sometimes under a magnifying glass, while the Consumer Confidence Report is almost completely unappreciated. In fact, most Americans don't have a clue that they receive an annual Consumer Confidence Report about how their drinking water is doing. When their annual report arrives, they open the envelope or email to make sure it's not important, and after a cursory glance, they toss it or delete it. A major reason for this lack of interest is that you basically need to have a translator on hand, preferably a chemist or water expert, to understand what all this stuff means.

For the very few who have the necessary background or are willing to put in that extra effort, this Consumer Confidence Report makes for quite interesting reading. We are, after all, talking about the ingredients in our drinking water. One of the tables lists the naturally occurring contaminants in your service area, so you might see arsenic or radium, at what levels they're allowed in your drinking water, and how your water provider is doing in keeping them out. After the Flint catastrophe, lead gets special attention these days. As one example, San Diego's report notes that 254 schools were sampled for lead. With few exceptions, your Consumer Confidence Report is about contaminants regulated by the EPA—which means that perchlorate won't appear unless it's regulated. Then there's a category called *disinfection byproducts*, which are taken very seriously down at your water treatment plant. And they have been for decades, because these were the first contaminants regulated by the EPA. It turns out this innocuous-sounding category was, and continues to be, one of the major challenges regarding drinking water safety. It all began with chlorination.

Chlorination of drinking water is widely hailed as *the* major health achievement of the twentieth century. Along with filtration, it has saved more lives than any other public health development in all of human history.[20] However, in the beginning, the idea of introducing a "poison" into the drinking water supply was highly controversial and had to overcome stiff resistance.

In 1902, Middelkerke, Belgium, installed the world's first chlorine disinfection system for drinking water.[21] Prior to that time, no municipality had ever added chemicals to their drinking water supply. Even filtration was considered unnatural by many. In 1908, Jersey City, New Jersey, joined the vanguard when it became the first city in the United States to chlorinate drinking water. Jersey City's history-making achievement came only after two high-profile court cases.

Located across the Hudson River from Lower Manhattan and with a backside view of the Statue of Liberty, Jersey City was the main port of entry for immigrants to the United States for decades as they arrived at Ellis Island. Like many other cities of the day, the city battled typhoid fever transmitted through unsanitary water. In 1904, Jersey City began receiving untreated water from a newly constructed reservoir. The contract had promised water that was "pure and wholesome," but it didn't turn out that way. Water with high concentrations of bacteria would periodically short-circuit the natural purifying processes of the reservoir. When the water company refused to install an expensive water filtration plant, the city sued. After a lengthy trial with testimony from some of the foremost engineers and public health experts of the day, the judge decided that the best course of action was to install sewers to capture and divert wastes away from the reservoir.[22]

Dr. John L. Leal, an advisor to the water company, had a much less expensive solution in mind. Leal's groundbreaking proposition was to add chlorine to the water as it left the reservoir. Leal was a physician and experimenter, and he knew that chlorine kills bacteria. Chlorine, the same poison used in gas warfare in World War I, would be added at levels low enough to maintain a safe water supply. The judge was skeptical but open-minded. He gave the water company three months to test Leal's plan and then report back.

In just ninety-nine days, George W. Fuller, the foremost American sanitary engineer at the time, designed the plant and supervised its construction. At a cost of only $5.60 per day, the plant produced water with very low levels of bacteria for Jersey City's entire water usage. After a second court trial with dueling expert witnesses, the judge ruled that chlorine disinfection was capable of supplying Jersey City with water that was "pure and wholesome." Neither sewers nor filtration were needed.

With the city's chlorination plant now in operation, the typhoid death rate in Jersey City almost immediately dropped in half and, ultimately, was driven to zero. The news traveled quickly, and other cities soon adopted the practice. Within a decade, chlorinating drinking water had spread to almost every large city in the country.

While this major public health breakthrough has saved countless lives, it's also an example of that old axiom—every solution creates new problems. In this case, it took over half a century to discover that chlorination creates suspected carcinogenic chemicals in drinking water. Unraveling this mystery required a combination of the development of sensitive instruments and innovative Dutch and U.S. chemists.

In the early 1970s, the Dutch chemist Johannes J. Rook worked for the Amstel brewery. His job entailed identifying the chemicals that were causing bad tastes in beer. Any good brewmaster will tell you that exceptional beer is all about chemistry, beginning with the chemistry of the water that you use. Rook discovered that he could detect the volatile (easily vaporized) chemicals responsible for undesirable flavors by sampling the air trapped under the bottle cap.[23]

When Rook began working for the Rotterdam Waterworks, his employers asked him to use a similar approach to measure the concentration of organic chemicals in the city's tap water. Rook discovered that chemical reactions between chlorine and dissolved humic substances in the water were causing high levels of chloroform, a suspected carcinogen. Humic substances form from the natural decay of wood, leaves, and algae, giving water rich in organic matter its characteristic yellow-brown color. By themselves they aren't toxic to humans.[24]

Around this same time, the link between chlorination and suspected carcinogenic organic compounds in drinking water was also being discovered in the United States. It started in New Orleans. For years, residents of New Orleans had been complaining about their drinking water having a chemical and oily taste. In 1972, the EPA concluded that the disagreeable taste was because of industrial wastes being discharged into the Mississippi River, and that chlorinating the drinking water may also play a role.[25] Outside the scientific community, little attention was paid to this finding until the summer of 1974, when a three-part series in *Consumer Reports* titled "Is the Water Safe to Drink?" got people's attention.

The real attention grabber came several months later, in back to back reports by the Environmental Defense Fund (EDF) and EPA. In the first report, a team of scientists led by the EDF reported statistical evidence that white males whose drinking water came from the Mississippi River had a 15 percent higher chance of dying from cancer than white males who consumed well water. The report noted that "the statistical analysis in this study is the first evidence in this country, to our knowledge, that carcinogens in drinking water are in sufficiently high concentrations to endanger human health."[26]

The very next day, the EPA released the preliminary results of a study showing that at least sixty-six synthetic (manmade) organic chemicals were present in Mississippi River water consumed by residents of New Orleans and nearby communities.[27] (The number of chemicals was later raised to ninety-four.) It was still debatable whether long-term exposure to the low concentrations of these chemicals could cause cancer, but most experts agreed that they shouldn't be in drinking water. To determine if the problem extended beyond New Orleans, the EPA launched a nationwide study of synthetic organic chemical contaminants in the drinking water of eighty cities.[28]

Another shoe fell a month later. In December 1974, EPA chemists reported evidence that some of the organic chemicals (chloroform and similar compounds) were being formed by chlorination.[29] These suspected carcinogenic compounds became collectively known as trihalomethanes—THMs for short. While it would have been premature to blow the whistle on THMs as a threat to public health, the evidence was piling up that adding chlorine to drinking water was producing chemicals believed to be responsible for increased rates of cancer.

Members of Congress took note of these rapidly escalating concerns about the safety of their constituency's drinking water. Legislation to set enforceable water-quality standards had languished in Congress for years. Virtually overnight, the one-two punch of the EDF and EPA studies changed everything. On December 16, 1974, President Gerald Ford signed the landmark Safe Drinking Water Act. The act authorizes the EPA to establish minimum standards to protect tap water and requires all owners or operators of public water systems to comply with these standards. The Safe Drinking Water Act is the primary law safeguarding the water we drink. It's also one of the most momentous challenges ever undertaken by the EPA.

Within a year after passage of the act, the EPA released results from its eighty-city survey of manmade organic compounds in drinking water. The news wasn't good. THMs were present in *all* chlorinated drinking water systems, and often at concentrations about a hundred times higher than any other organic chemical.[30]

When the EPA issued its first drinking water regulations in 1975, as a stop-gap action the agency simply adopted U.S. Public Health Service standards for twenty-two contaminants. THMs and synthetic organic chemicals (other than six pesticides) were not included.[31] Four years later, the EPA established its first drinking water regulation—a standard of one hundred parts per billion for the total concentration of four THMs. According to the EPA, a person who drank water containing this concentration for their entire life had an increased risk of developing cancer of four in ten thousand.[32]

The EPA was facing a major catch-22. No one wanted to return to the conditions of the early 1900s, when deadly waterborne diseases like cholera, typhoid fever, and diarrheal diseases were common—and life expectancy was forty-seven years. The problem was how to maintain pathogen-free drinking water while minimizing the risks from disinfection byproducts. The search for solutions took several twists and turns and continues to this day.

Since the late 1970s, over six hundred disinfection byproducts have been found in chlorinated tap water.[33] Among these is a family of compounds known as haloacetic acids. These are fairly simple chemicals consisting of a molecule of acetic acid (the main acid in vinegar) with one or more hydrogen atoms replaced by chlorine or bromine. Haloacetic acids are more toxic than THMs, but they still can't explain the carcinogenicity of chlorinated drinking water.[34] Disinfection byproducts have been primarily linked to bladder cancer but the association remains unclear to this day.[35] With all these complications, the EPA has focused on THMs and haloacetic acids as "indicators" of disinfection byproduct toxicity.

Considerable effort has gone into figuring out how to structure water treatment processes to minimize exposure to disinfection byproducts. One solution is to head the whole problem off at the pass by treating drinking water with activated carbon to remove humic substances *before* chlorination. When the EPA issued the total THM rule, it also proposed requiring water treatment facilities in large cities that used surface water to use granular activated carbon. The proposal was quickly shot down by the water industry as too complex and expensive.[36]

Fortunately, there are other ways to outfox (or at least reduce) chlorination's troubling byproducts. One method is to treat the water with ozone, a powerful oxidant. Ozonation avoids the taste imparted by chlorination and is popular in Europe and Asia, but it's expensive. Ozone also degrades rapidly, which means that utilities often add chlorine after ozone treatment to prevent water from becoming contaminated during transport to homes and businesses. Ozonation creates its own carcinogenic compound—bromate.

Another approach to minimizing disinfection byproducts is to use a less reactive form of chlorine known as chloramines. At first, this alternative looked like a win-win—it's relatively inexpensive and easy to implement. Many water utilities in the United States switched over to chloramines. But the honeymoon was short-lived when it was discovered that chloramines produce their own toxic byproducts. Among them is NDMA, a highly toxic chemical to the liver and a proven carcinogen. Chloramines also have been linked to elevated levels of lead in tap water.[37]

After more than four decades of major effort and expense, the EPA and public utilities are still wrestling with how to assure pathogen-free drinking water while minimizing the dangers of disinfection byproducts. The risks of microbial disease continue to be the most significant public health concern. Even with all of today's sophisticated water treatment, community water systems cause approximately sixteen million cases each year of acute gastroenteritis, with symptoms including diarrhea, vomiting, and fever.[38] Meanwhile, THMs and haloacetic acids have been greatly reduced since the EPA first set standards, yet disinfection byproducts continue to comprise the largest percentage (about 30 percent) of drinking water violations in community water systems nationwide.[39]

Disinfection byproducts were the first contaminants to be regulated in drinking water by the EPA. Perchlorate is the most recent contaminant planned for regulation. Both illustrate how setting a drinking water standard is a highly complex and multi-faceted undertaking. An initial hurdle is to develop laboratory methods that can measure the contaminant at the levels of concern—often in the few parts per billion range. (One part per billion is like a drop of water in an Olympic-sized swimming pool.) EPA scientists must then estimate human exposure to the contaminant of concern, which is a time-consuming but relatively straightforward step. Samples are collected and analyzed from a representative set of drinking water systems and the results interpreted. Then comes the hard part. Conducting a risk analysis that has any chance of weathering pushback down the road depends on extensive evaluation of toxicological and long-term epidemiological studies for a host of potential health problems. The results of these risk analyses are inherently controversial: What level of chemical exposure increases the risk of disease? Do exposures at certain ages, such as infancy or during pregnancy, have more severe consequences? How does exposure through drinking water relate to accumulation in the body? Adding to the complexities, risk analyses must often consider other exposure pathways, such as food. And after all that comes assessing the economic feasibility of water treatment to reduce levels of the contaminant in drinking water. Finally, the costs and benefits of any proposed regulation must be compared, leading to yet another highly controversial topic—quantifying the benefits from saved lives and reduced illnesses. Accompanying all these challenges can be (and often is) intense pushback from affected industries. In other words, what the EPA has to accomplish to set a drinking water standard is, in many ways, comparable to climbing K2.

Twelve years after the Safe Drinking Water Act was passed, the EPA had set a drinking water standard for only one contaminant— total THMs. Frustrated with the slow pace, Congress turned up the heat in 1986 with amendments to the Safe Drinking Water Act. Under the new rules, the EPA was required to set standards for eighty-three specified contaminants, and Congress gave them three years to do it. On top of this high-pressure plan, the EPA was required to set standards for twenty-five additional contaminants every three years thereafter. Initially, the pace picked up, and the EPA set standards for eight volatile organic compounds in the first year. But, as we've just seen, there was no way such a pace could continue indefinitely. Another thing Congress hadn't considered was that even if the EPA achieved this goal, it would lead to astronomical costs for water utilities—which means the ratepayers.

In 1996, Congress passed amendments that laid out a new procedure for selecting contaminants to regulate. This involved a three-step process.

First, every five years, the EPA must issue a Contaminant Candidate List of unregulated contaminants that are of concern in public water supplies. The most recent one, issued in 2016, listed ninety-seven chemicals and twelve microbial contaminants. (Perchlorate was included on the first three lists.)

Second, the EPA must target up to thirty unregulated contaminants that large public water systems and a representative sample of small systems need to monitor. Again, the EPA updates this list every five years. The data collected from this nationwide monitoring allow the EPA to estimate how many people are exposed to a specific contaminant and at what levels. This step, known as the Unregulated Contaminant Monitoring Rule, has a key role in determining the need for regulation. (Perchlorate was included among the first thirty unregulated contaminants. Cyanotoxins are included in the current list.)

Finally, the EPA must make "regulatory determinations" on at least five contaminants every five years. What this basically involves is deciding if the agency is going to regulate a contaminant or let it off the hook. If and when the EPA decides to go the distance and regulate a contaminant, the law requires a subjective decision by the EPA administrator. Congress's exact wording is: "in the sole judgment of the Administrator" there is "a meaningful opportunity for health risk reduction." Putting the onus on the administrator makes sense when the person at the helm is dedicated to the agency's mission. In the case of someone like Scott Pruitt, it's open to tremendous abuse.

Once the decision has been made to regulate a contaminant, the EPA then needs to determine two standards. The first one, called a Maximum Contaminant Level Goal (MCLG), is just that—a *goal*, meaning it's non-enforceable. In addition, no consideration is given to cost or feasibility of treatment. MCLGs are set at zero for carcinogens. The enforceable standard, the Maximum Contaminant Level, is set as close as possible to the MCLG, but takes costs and benefits of regulation into consideration. Throughout this entire process, the EPA is usually bucking opposition from powerful industries who want less strict Maximum Contaminant Levels.

It's not just drinking water standards that are of concern. Since 1996, the EPA has provided over twenty billion dollars in loans to communities to improve their drinking water facilities through the

Drinking Water State Revolving Fund.[40] The America's Water Infrastructure Act of 2018 increased the authorized amounts. Despite these substantial investments, the EPA estimates that water utilities nationwide will need to invest $472 billion over the next two decades to meet the growing challenges of ensuring safe tap water.[41] That's a big gulp.

Over the past four decades, the EPA has set drinking water standards for about one hundred chemical and microbial contaminants. Other contaminants are slowly working their way through the three-step process. Among these are metals like chromium-6, various pesticides and industrial organic compounds, cyanotoxins from harmful algal blooms, individual disinfection byproducts, and a suite of "forever chemicals" known as PFAS (that we'll turn to in chapter 10), yet these barely scratch the surface of the vast array of chemicals showing up in drinking water—personal care products, pharmaceuticals, hormones, new pesticides, and industrial chemicals. The list seems virtually endless, and there's no consensus on the dangers all these contaminants may pose to our health. The bottom line is that it's impossible to try to figure out what to do about every possible contaminant. Which brings us to a basic commonsense question— why not invest more effort into protecting drinking water sources? Or as Benjamin Franklin aptly advised, "An ounce of prevention is worth a pound of cure."

The idea of protecting drinking water sources didn't get much attention until 1986, when amendments to the Safe Drinking Water Act addressed areas that recharge public water supply wells.[42] Groundwater provides drinking water for more than one hundred million people in the United States.[43] Most people know not to dump a poison or pesticide on the bank of a stream, because it will work its way into the water. However, there's much less awareness about groundwater, both in its scope and in its vulnerability to contamination. Unlike streams, groundwater is pretty much everywhere, and much of what goes on the ground sooner or later works its way into this critical water resource.

When the idea of protecting drinking water wells became part of the Safe Drinking Water Act, the strategy was fairly simple. First, delineate the area that contributes water to each public supply well (defined as the wellhead protection area). Second, identify potential pollution sources in those areas. Third, raise awareness of the vulnerability of public supply wells to contamination. In the subsequent 1996 amendments to the Safe Drinking Water Act, this simple but challenging idea of protecting source water was expanded to include surface-water sources—streams, lakes, and so forth. Neither of these amendments, however, mandated any further action, and funding dropped, so the whole thing has languished—with some notable exceptions. [44]

Indianapolis, as well as the whole of Marion County in which it lies, has an exemplary groundwater protection program. The county relies on groundwater for more than 25 percent of its drinking water. Groundwater is also critical for future growth, and Indianapolis is a city that wants to grow. Therefore, protecting their groundwater is crucial. Most of Marion County's productive aquifers are shallow, making them highly vulnerable to contamination. As an old industrial city, Indianapolis has a long history of contaminating its groundwater. The White River runs right through the city and is also polluted. So city leaders decided to get serious.

The wellhead protection areas for each of Marion County's seven public supply wellfields were mapped. [45] Within this framework, about a thousand commercial and industrial sites have been identified as potential contaminant sources due to their business practices. These include everything from gas stations, auto repair shops, dry cleaners, and mortuaries (with their substantial use of embalming fluids that you definitely don't want in your water supply), to the full gamut of old rusting industrial operations and mom-and-pop manufacturing emporiums. With its wellhead protection areas mapped and the list of potential contaminant sources in hand, as far as the Safe Drinking Water Act amendments go, the county could have declared success. However, Marion County was just getting warmed up.

Since 1996, any company seeking a building permit within a wellhead protection area has had to clear a number of hurdles. After completing the usual paperwork, the permit undergoes scrutiny by a technically qualified person to ensure that the groundwater will be protected. This includes such things as leak-proof containment areas for chemical storage and a proper chemical spill kit available at all times—and employees knowing how to use it. But the much more challenging problem was what to do about all those old leaking industrial operations that have been there for decades, as no one has ever dared to come knocking and tell them how they should be doing things. To avoid such predictable pushback, the county adopted a voluntary approach. Businesses within a wellhead protection area were encouraged to contact the Marion County Wellfield Education Corporation (MCWEC), a not-for-profit group funded, in part, by water use fees. They would then receive a free (and strictly confidential) business assessment by a trained environmental consultant.

Haley Waldkoetter worked as one of the county's trained environmental consultants. "At first," she says, "people were suspicious that it was a scam. Some of them suspected that I worked for immigration. But once we redesigned our website, that helped a lot." With the White River running right through the city, most people had no idea that groundwater had anything to do with their water supply. Consequently, MCWEC's theme became, "Groundwater is your drinking water! Protect it!" and Waldkoetter's main job was education.

She also received one. "When I showed up at an old warehouse, I'd often be escorted through huge stacks of discarded chemical containers and rows of rusted metal drums leaking oil onto a cracked concrete floor. There'd be shelves stuffed with cracked plastic bottles and improperly sealed cans tattooed with skull-and-crossbones warnings and how-to-handle labels. Floor drains often discharged directly into the ground." In other words, what Waldkoetter found was a whole range of source water protection nightmares.

Waldkoetter, however, is a very nice young woman and people could see that. She also had free items to offer. Among the favorites were secondary containment equipment and spill clean-up kits. It was all simple stuff, like a large pan to put under your tank of waste oil. "The amazing thing," Waldkoetter says, "is that when we'd go back months, or even years later, these things were still being used. Nobody had any idea what it was there for, but they were using it. And so, we started labeling these things with our name and contact information. That way, if they or a new owner had a question, they could contact us."

MCWEC had an excellent business-friendly approach, but as years passed the group realized they weren't getting the job done. The problem was their program was all carrot, no stick. "Years after businesses received their building permits, we'd find major non-compliance issues. And the businesses that needed help the most were the least likely to request it."

Waldkoetter worked for John Mundell, who has been involved with Marion County's source water protection program from the beginning. "Historically," John explains, "less than 10 percent of the identified potential chemical sources in the wellfields have worked with MCWEC in any significant way. And because the program was anonymous and voluntary, any dangers to the wellfield that were discovered could not be reported." It became increasingly obvious that a more aggressive approach was needed, and so the county public health department got on board.

In 2017, the health department enacted a mandatory and enforceable Wellfield Health Code. Under this new code, all businesses within wellhead protection areas are required to use best management practices, such as employee training on spill response and prevention, proper container labeling and storage, and secondary containment for hazardous materials. To ensure this is being done, health department inspectors visit each business and then return at suitable intervals. Depending on the level of concern, repeat violators receive notices of violation, fines, and possibly even legal action. During the first visit, the health inspector hands out a card with

MCWEC's contact information and suggests they contact them for advice. The phone is ringing these days.

Waldkoetter has gone on to grad school and Rachel Walker now has her job under the new regime. Walker sits down with the business owner and works out a manageable step-by-step process. She has the contacts and solutions at her fingertips, such as who to call to have old oil removed, or where to go for other problems. If Walker sees meaningful progress during her return visit, she lets the health department know that they're making a good-faith effort. There's a lot of flexibility, because everyone knows these things take time. After all, they're not out to bust people. "Our job is to protect the county's water supply," she emphasizes.

No discussion about drinking water would be complete without including today's bottled water craze. A number of books and articles on this topic have created public awareness about the problems of bottled water, and so for our purposes, a brief overview will do.

Many people drink bottled water because they believe it's safer than tap water. This was definitely true for those living through the Flint, Michigan, crisis and continues to be the only safe option for many disadvantaged communities. However, for the general population, this is simply not true. Many popular brands, the ones you see rolling out of stores by the caseload, are just bottled tap water.

Bottled water is a very expensive way to drink tap water, costing a thousand times (or more) per gallon than just filling your glass at the tap. It's also expensive in terms of resources. The energy used to produce the plastic bottles for U.S. consumption is equivalent to fueling more than a million cars and light trucks for a year.[46] In addition, water weighs over eight pounds per gallon, which translates into large energy costs for transporting bottled water, particularly from Europe and the South Pacific.

For the average American, buying bottled water for their primary source of drinking water makes little sense. If you're buying one of the cheaper brands, it makes no sense.

The Safe Drinking Water Act was the brainchild of an idealistic Congress in a period of supercharged environmental activism. Society was demanding a sweeping environmental cleanup, a swabbing of the ship from helm to hold—and so things got done. Progress has subsequently slowed. A recent national study found that 9 percent of community water systems in the United States, affecting nearly twenty-one million people, violated water-quality standards at least part of the time.[47] A single violation doesn't necessarily mean the water is unsafe, but it is a red flag indicating a need for improvement.

The EPA must simultaneously navigate through myriad political crosswinds while addressing complex science and risk assessment issues. Virtually every contaminant in question has powerful forces aligned against regulation. Drinking water standards often become minimum clean-up standards for Superfund sites, which means that companies and government agencies (such as the Department of Defense in the case of perchlorate) are on the hook for cleaning it up. It's no surprise that they have a vested interest in blocking new regulations. This much is obvious. What is less obvious, and where people don't tend to connect the dots, is how regulating a chemical may translate into a huge cost to water utilities that they pass off to the ratepayers. There's no surer way to pack a city council meeting—including along the sides and back, down the hall, and even filling a second room where the meeting is televised—than having a proposed water rate hike on the agenda.

Compared to much of the world, the United States has good quality drinking water. The glass is truly much more than half full. Yet continuing (and ideally improving) this standard into the future requires a joint effort among the EPA, state public health agencies, water utilities, proactive communities such as Indianapolis in Marion County, and every single one of us who make daily choices about how we use fertilizers and pesticides on our lawns and gardens, and how we dispose of chemicals. There was a time when it took a village to protect its drinking water. It now takes an entire nation.

3

ENVIRONMENTAL JUSTICE

It's regular, good, pure drinking water, and it's right in our backyard.
—Mayor Dayne Walling's toast, as he drank the first cups of
water from the Flint River[1]

Flint, Michigan, was once a thriving city built around the automobile industry. General Motors was founded in Flint in 1908. The city
boasted the highest average income and lowest unemployment rate
in the nation. Beginning in the late 1960s, Flint followed the pattern
of other Rust Belt cities. It lost half its population and many of those
remaining were unemployed, with more than 40 percent of the city's
mostly black population living below the poverty line. Crime rates
became among the worst in the nation. The city also went bankrupt.
In an attempt to bring Flint's finances back in order, Michigan
Governor Rick Snyder appointed an emergency manager to take
over financial operations from the mayor and city council. This
decision proved to be an ill-fated step leading to a fixation on money and balance sheets at the expense of public safety and health. As
Michigan Attorney General Bill Schuette would later charge, "It's
all about numbers over people, money over health."[2]

For decades, Flint had purchased its water from Detroit's utility,
which was piped seventy miles from Lake Huron—an arrangement

that worked well. Then along came Flint's state-appointed emergency manager, who became convinced they could save money by switching to another water authority that was being formed to tap into Lake Huron. However, there were two impediments. First, Flint would be responsible for 35 percent of the cost of the new pipeline, which would be financed by issuing bonds. Yet Flint literally had a zero-credit rating. This impediment was circumvented by a complicated bond scam, later described by criminal prosecutors as a "sham transaction designed under false pretenses."[3] The second problem involved finding a temporary water supply while the new pipeline was being constructed. In 2013, Flint's emergency manager approved a plan to temporarily change the city's water source to the Flint River. With the push of a button, on April 25, 2014, the city stopped buying treated water from Detroit and began drawing water from its own historically polluted river and cleaning it with a hastily refurbished old treatment plant.

Almost instantaneously, residents began complaining about the taste, odor, and appearance of the water. People reported rashes and welts on their bodies. Their hair started falling out. In these early days, no one knew about the lead problem, but a cascading series of other water quality problems soon appeared. In September, the city issued a boil water advisory in response to coliform bacteria detected in the water. To solve the bacteria problem, chlorine levels in the water were increased. This caused another problem. The chlorine reacted with the organic matter in the water, resulting in levels of disinfection byproducts (trihalomethanes) above drinking water standards. The chlorine also made the water corrosive, rusting machinery and parts at the General Motors (GM) engine plant. GM switched back to Detroit water. No one, however, made the connection that if the water was acidic enough to corrode machinery, what effect could it be having on Flint's old lead pipes carrying water to people's homes?

By early 2015, angry residents were showing up at public meetings holding up bottles of brownish, foul-smelling water they had drawn from their taps (and still unaware of the lead problem). Detroit offered to reconnect Flint to its water system. The city council voted in support of the idea, but the state-appointed emergency manager refused, insisting that the water was safe.[4]

The discovery of Flint's lead contamination and the battle for official recognition of this problem would require dogged determination by many people in the community. Among these were a mother, a regulator for the U.S. Environmental Protection Agency (EPA), a scientist, and a pediatrician. These four (along with others) confronted a chorus of deniers at the city and state level. In addition, the regional EPA office failed to do its job and intervene.

Lead is a potent neurotoxin. Children are particularly vulnerable because of their rapidly developing brains and nervous systems. Exposure to lead in the womb or at a young age can result in lowered IQ and behavioral changes, such as shortened attention span and increased antisocial behavior. Lead exposure in adults can cause kidney problems and high blood pressure. There is no known level of lead exposure that is considered safe.

The unfolding of the lead story began with LeeAnne Walters, a mother worried about her family's recent health problems. Her children were breaking out in rashes and suffering from other ailments. One of her children, who had a weak immune system, was losing weight and having occasional problems pronouncing words. In early 2015, Walters asked the city to test her tap water. When the results came back, a city employee left a voicemail that her water had dangerously high levels of lead and to stop drinking it. City water officials blamed Walters's house plumbing for the problem and hooked up a garden hose running from her neighbor's house.[5] Officials insisted they were regularly testing the water and there were no lead problems in Flint's water.[6]

Walters started examining Flint's water quality reports. Trained as a medical assistant, she discovered that Flint's water was more corrosive than Detroit's. She also was concerned that the city employee who tested her water had run the faucet for several minutes before taking the sample. This would flush out the lead that had been leached from her pipes, lowering the sample results. Walters shared the list of chemicals that the Flint treatment plant was using

with Miquel Del Toral, an EPA drinking water expert in the regional office.

Del Toral immediately recognized that orthophosphate (a corrosion inhibitor) was not on the list. As a result, Flint's water, which
he later described as "corrosive as hell," would quickly dissolve
lead from houses with lead service lines—the pipes that connect
individual homes to the water mains in the street.[7] He sent an email
to his EPA colleagues saying that they needed to investigate Flint's
lack of corrosion control.

With state and local officials still claiming that everything was
fine, Del Toral sent a memo to his bosses at the EPA regional office.
He explained that Flint's failure to use corrosion controls had created a major public health concern. He also alerted them to the practice of pre-flushing before collecting samples. He shared the memo
with LeeAnne Walters, who passed it along to the local office of the
American Civil Liberties Union (ACLU). The ACLU published Del
Toral's memo.

The reaction of the Michigan Department of Environmental
Quality (DEQ) followed a pattern that would continue as the evi-
dence of lead contamination unfolded—that of knee-jerk denial.
The DEQ spokesman denounced Del Toral as "a rogue employee."[8]
In words he would later regret, the spokesman stated emphatically:
"Anyone who is concerned about lead in the drinking water in Flint
can relax."[9] Meanwhile, instead of acting to protect the public based
on Del Toral's concerns, the regional EPA office reprimanded him.

Del Toral put Walters in touch with Marc Edwards, a Virginia
Tech scientist who was an expert on corrosion of water pipes. Instead of taking a single sample after "pre-flushing," as the city had
done, Edwards instructed Walters to collect a series of samples from
the water that had been in her pipes long enough to reflect the levels
her family was exposed to under normal use. Edwards found higher
levels of lead in these samples than those previously reported by the
city. In one sample, lead levels were so high that the water qualified
as hazardous waste.[10]

In the early 2000s, Marc Edwards had exposed lead contamination in the water supply of an economically depressed part of Wash-

ington, DC. For over a decade, he battled the EPA and Centers for Disease Control and Prevention over the issue. Despite challenges to his credentials, Edwards refused to back down. He was eventually vindicated by a Congressional study. Marc Edwards is a formidable and media-savvy activist—or as he likes to put it, "a troublemaker."[11]

During the summer, Edwards and a team of volunteer students drove over five hundred miles from Blacksburg, Virginia, to Flint, Michigan, to collect hundreds of drinking water samples. Edwards claimed this was "the most thorough independent evaluation of water in U.S. history."[12] Back at Virginia Tech, they ran the same tests that Flint officials said they had performed and found levels of lead that were clearly in violation of EPA standards. In September, with LeeAnne Walters by his side and encircled by activists, Edwards announced his findings at a news conference on the lawn of Flint's City Hall. "The levels that we have seen in Flint are some of the worst that I have seen in more than 25 years working in the field," he later told Michigan Radio.[13]

The DEQ was again dismissive. The department's spokesman explained to a local journalist that Edwards and his team had "only just arrived in town and quickly proven the theory they set out to prove."[14] Two weeks later a local pediatrician blew the lid off the Flint lead story, making it almost daily national news.

Mona Hanna-Attisha, a first-generation Iraqi immigrant whose parents fled Saddam Hussein's murderous regime, is a pediatrician who also has a master's degree in public health. Hanna-Attisha had become concerned after hearing about Del Toral's memo from a friend. She began to investigate, gathering records of lead levels in blood for all the children who had been tested at her hospital. Comparing lead levels before and after the switch to Flint River water, she was stunned to find that the percentage of children with high lead levels in their blood had doubled in many areas, and even tripled in some parts of the city. On September 24, 2015, Hanna-Attisha announced her findings at a press conference.[15]

The next day, Flint officials issued a very cautious lead advisory to residents: "While the City is in full compliance with the federal

Safe Drinking Water Act, this information is being shared as part of a public awareness campaign to ensure that everyone takes note that no level of lead is considered safe."[16] At the same time, state regulators publicly denounced Hanna-Attisha's findings. The DEQ spokesman said the water controversy was becoming "near-hysteria." The Michigan Department of Health and Human Services also questioned the pediatrician's results, saying that they were not seeing those numbers in their larger data set.[17]

The state soon changed its tune. Less than a week after Hanna-Attisha's press conference, Governor Rick Snyder pledged to act. Two weeks later, after eighteen months on corrosive Flint River water, the state announced it was changing the source of the city's drinking water back to Lake Huron. The governor's office downplayed the significance of this decision, stating that the Detroit water "will be easier to manage. It comes from a more stable source than the river, it is fully optimized for corrosion control, and it is clear that residents of Flint have more confidence in this water source."[18]

City water officials began adding phosphates to the water to try to rebuild a protective coating inside the pipes, but the corrosion damage in the pipes could not be quickly undone. State and city programs began to provide city residents with water filters and bottled water. In January 2016, in an illustration of the bizarre circumstances, Flint residents were invited to bring their children to a local elementary school for a "Lead Testing and Family Fun Night," combining a school carnival with medical tests to check children's blood.[19]

The excuses and blame game unfolded with fingers pointing in all directions—at state and city officials, the governor, and the EPA. DEQ Director Dan Wyant explained the lack of corrosion controls as a misunderstanding: "It's increasingly clear there was confusion here, but it also is increasingly clear that DEQ staff believed they were using the proper federal protocol here and they were not."[20] City and state officials argued that they had operated under the mistaken belief that they were not required to treat the Flint River water for corrosion until after two six-month monitoring periods.

During at least the previous six months, EPA officials at the regional office had been battling with the DEQ about the need for corrosion control and how to apply its rules for lead sampling. Instead of moving quickly to take preventative measures, EPA officials tried to coax Michigan's DEQ to take action.[21]

The disaster in Flint, Michigan, is a failure of a fundamental precept upon which the EPA was formed. The entire EPA system is dependent upon the regional offices maintaining sufficient independence from the states they oversee. Flint is a tragic example of their failure to do so. The regional office was completely lacking in a sense of urgency to act and failed to intervene despite clear warnings from its own employee and others about a serious health risk to Flint residents.

At the end of 2015, the DEQ director and spokesman both resigned. Soon thereafter, the EPA regional director was forced out. Governor Snyder transferred power from the emergency manager back to the city. In January 2016, the governor and President Obama declared a state of emergency in Flint and police officers began delivering cases of water, lead testing kits, and filters to homes. The EPA took over the lead monitoring. Congressional hearings and lawsuits soon followed.[22] Michigan's attorney general filed criminal charges against fifteen local and state officials and water system operators. Included were charges of involuntary manslaughter against five officials for having failed to notify the public or act on a second serious health problem affecting Flint—Legionnaires' disease.[23]

During 2014 and 2015, the county that includes Flint experienced the third largest outbreak of Legionnaires' disease in U.S. history. At least eighty-seven people were infected and twelve died. Legionnaires' disease is named for the first recognized case, when 182 attendees were infected and twenty-nine died at an American Legion convention held in Philadelphia, Pennsylvania, during the 1976 bicentennial. The disease is a virulent form of pneumonia that grows in plumbing systems and is usually spread through breathing mist in the air. People can become exposed from water in fountains, hot tubs, showers, or cooling systems. It is most harmful to the elderly and those with weak immune systems. The connection between Flint's water and the *Legionella* outbreak remains inconclusive.[24] Not debatable, however, is that officials were aware of the outbreak and failed to act. Despite a wave of such cases, no public warning was issued until early 2016.

In April 2018, the state of Michigan announced that Flint's water had met lead standards for about two years and the free bottled water program would end. By this time, sixty-two hundred lead service lines had been replaced, about a third of the way toward planned replacement of all lead service lines in the city by 2020. However, confidence in authorities was still shaken, and people lined up outside water distribution points to load up on the last of the free bottles.[25]

Flint brought lead in drinking water to national attention, including a basic unresolved question. Should all lead service lines nationwide, serving an estimated six to ten million homes, be replaced? And if so, who pays? With replacement costs averaging around five thousand dollars per line, eliminating all lead service lines would run into tens of billions of dollars. An additional impediment is that once a line runs underneath a homeowner's property, it belongs to the owner. Low-income neighborhoods disproportionately tend to have more lead service lines and are the least able to afford replacement. Municipalities and utilities argue that the problem can be avoided in many cases by properly treating the water with corrosion inhibiters.

It is widely recognized that major changes are needed to the EPA's Lead and Copper Rule. The rule was promulgated in 1991 to minimize lead and copper levels in drinking water, primarily through requirements for corrosion control. In addition to questions surrounding lead service line replacement and corrosion control requirements, key issues include where, when, and how to sample for lead in homes, schools, and childcare centers. A long overdue revised rule is scheduled for release in 2019.

For the people of Flint, Michigan, distrust of public officials remains high. Fortunately, follow-on studies have found that changes in blood lead levels of young children in Flint were relatively modest compared to the days of leaded paint and gasoline. [26] Nonetheless, the neurological effects of lead are considered to be irreversible. United Way estimated that six to twelve thousand children may have been exposed to unsafe lead levels. [27] The tragic irony of all this suffering is that a preventative solution would have been maddeningly simple, and cheap. For only two hundred dollars a day, Flint's pipes could have been protected by adding common anticorrosion chemicals. [28] It also is lost on no one that what was allowed to happen in Flint never would have occurred in a white, middle-class community.

Lead is not the only environmental justice problem associated with drinking water. Many drinking water violations involve small systems in rural areas that don't have the capital to afford treatment costs and maintain proper equipment and trained personnel. The EPA can help by targeting grants for improvements to these systems, but operation and maintenance costs require more durable funding. This problem has come to a head in California's agricultural communities, which are overwhelmingly Latino with high poverty rates and virtually no political voice.

Among the worst hit areas is California's agricultural heartland—the San Joaquin Valley. About 185,000 valley residents are served by water systems that fail to meet drinking water standards because of nitrate and pesticide contamination, as well as naturally occurring arsenic and uranium. [29] What this means is that residents pay the triple penalty of increased health risks, higher water bills, and having to purchase bottled water. The people most affected are the least able to afford the extra costs, which are considerable. Many of these families spend up to 10 percent of their meager income on buying water at local stores or water vending machines. There are now state-driven efforts to reduce agricultural nitrate loads to groundwater, but any real impact on drinking water quality will take decades. For many, if not most, disadvantaged communities, the only effective solution is to upgrade water treatment.

In 2006, attorney Laurel Firestone teamed up with community organizer Susana de Anda to start the Community Water Center, a nonprofit environmental justice organization. The center's goal is for all California communities to have access to safe, clean, and affordable water. The group describes itself as a "catalyst for community-driven water solutions through organizing, education, and advocacy." If you visit the Community Water Center's main office in Visalia, you can't help but be struck by the hard-working and no-frills office environment, reminiscent of the civil rights movement of the 1960s. This time, the focus is on safe drinking water as a basic human right.

It's been an uphill battle, yet the Community Water Center has made steady, and impressive, progress. Since opening the doors in 2006, they have worked with over eighty California communities to improve their access to safe and affordable water. They have trained thousands of residents as "clean water advocates." They have provided technical and legal assistance to over fifty local water boards and organizations that are struggling with how to manage water systems. And through community mobilization and intensive lobbying, the Community Water Center has put considerable pressure on the state to help struggling communities obtain safe drinking water.

A major achievement came in 2012, when the state legislature passed a Human Right to Water law. Patterned after the 2010 UN General Assembly's resolution of the same name, California became the first state in the country to recognize that every human being has the right to "safe, clean, affordable, and accessible water adequate for human consumption, cooking, and sanitary purposes." This all looks good on paper, but there's a major limitation—state agencies are merely *encouraged* to *consider* this policy in their work. In addition, the legislation didn't appropriate money or levy any taxes to actually make it happen. Some funding became available in 2014, when voters approved a ballot measure for water projects. In 2015, the legislature passed a bill giving the state authority to help small, struggling water systems by encouraging bigger neighboring systems to consolidate activities with them, *if possible*.

In 2017, Governor Jerry Brown proposed a statewide tax on drinking water to fix problem wells and treatment systems serving small and disadvantaged communities. The tax would increase residential water bills statewide by ninety-five cents per month. Low-income earners would be exempt. Businesses would pay up to ten dollars a month. Agriculture would also contribute through a tax on fertilizer manufacturers and dairy producers. This proposal was supported by an unlikely alliance of farmers (who, in return, got some relief on enforcement of regulations on nitrate contamination) and environmental justice advocates. It was opposed by urban water utilities concerned about the precedent of setting fees on urban water users for underfunded water problems and by skittish lawmakers hesitant to add a new tax. Despite two years of continual and passionate effort by the Community Water Center and others, the Brown administration failed to pass the legislation.

In 2019, incoming California Governor Gavin Newsom highlighted safe drinking water for disadvantaged communities as a priority and announced his support for the tax scheme. Meanwhile, the water utilities, under pressure to deal with this black eye on their industry, supported alternative legislation focused on small system governance and consolidation along with a fund to help bring small water systems into compliance. In July 2019, thirteen years after the Community Water Center was founded, the state committed to spending $130 million a year for ten years to help distressed water systems.

In addition to the public systems, many people in the San Joaquin Valley rely on private wells, which are not subject to federal drinking water regulations. These private well owners are on their own for testing and water treatment. To help this segment of the population, the Community Water Center has conducted free water testing in much of the valley for private well owners. They followed up with educational materials about the test results and home water treatment options. While this is a start, many people with private wells need financial assistance to actually solve the problem.

Working for environmental justice in finding a solution to serious drinking water problems requires persistence, political savvy, and the patience of Job. A mother, a pediatrician, and unsung community organizers in Flint, Michigan; the Community Water Center in California; and many other committed advocates across the country have these qualities in droves.

Part II

Water Pollution

4

A WICKED PROBLEM

A wicked problem is so complex socially that it has no solution—it can only get better or worse.

When Frank Perdue inherited his dad's chicken farm in 1952, it was bringing in six million dollars a year. Most farmers of that time would have been in chicken heaven, but Frank viewed the farm's success as a work in progress. By following his three lodestars— product, production, and profit—he began a quest for absolute efficiency in animal production in the same way Henry Ford revolutionized building a car. Perdue often worked eighteen-hour days and slept on a cot in his office. Every component of every part of the operation was broken down to its smallest parts and analyzed, from the moment a chick broke out of its shell to the plump saran- wrapped broiler in the supermarket case—proudly displaying the Perdue label that meant *quality*. "My chickens eat better than you do," was one of Frank's favorite advertising slogans that he'd deliv- er in that iconic, down-home twang. "The only way to eat as good as my hens is to eat my hens!" This balding scrawny guy, who curious- ly resembled a plucked chicken, appeared in over two hundred tele- vision ads. Anyone could see that Frank wasn't some slick actor working for a company where all they care about is money. *No sir!* Frank Perdue was the genuine article, someone you could trust. By

the 1970s, he was shipping more than two million broilers a week—
packed in ice, not frozen. "Freeze my chickens?" Frank would
screech at the cameras. "I'd rather eat beef!" When he died in 2005,
the company had sales of $2.8 billion and was selling more than
forty-eight million pounds of chicken products and nearly four million pounds of turkey products *a week*. Perdue Farms employed
nearly twenty thousand people. Frank Perdue had pioneered the
consummate animal factory.[1]

Perdue steadily expanded the original family farm into an operation that covered large swaths of the Delmarva Peninsula, com-
prised of Delaware and the eastern shores of Maryland and Virginia.
Much of this region is a rural holdout of poor shantytowns where
labor is dirt cheap. Harriet Tubman was from the Delmarva. Her
thirteen return trips to rescue relatives and friends through the
Underground Railroad encompassed the same terrain as the all-but
slavery in Perdue Farms factories. Frank Perdue viewed the human
component of his operation as just another cog in the wheel that
needed to be tuned to maximum efficiency. Workers stood on a
concrete floor at the conveyor belt, robotically moving an arm up
and down, or out and back, opening and closing the same hand
thousands of times each day. There were few rest periods, sick days,
or days off. If you couldn't cut it, you were gone. It all came down
to Frank Perdue's favorite advertising slogan: "It takes a tough man
to make a tender chicken." As locals increasingly became unwilling
to work under the inhumane conditions, they were replaced by immigrants desperate for work.

The methods developed by Frank Perdue for raising and processing chickens are now commonplace throughout the poultry industry
(although modified some after considerable pressure from animal
welfare groups).[2] Tens of thousands of chickens are crammed in
sheds where the air is almost unbreathable from chicken manure's
toxic ammonia fumes. Lights burn almost twenty-four hours a day
to keep the birds eating nonstop. After seven weeks, the birds have
grown to market weight (it used to take three times as long, back
when chickens got to be chickens) and are transported to processing
plants. The birds routinely suffer broken bones because they're bred

to be top-heavy, and because workers grab them by the legs and slam them into transportation crates. When the chickens arrive at the slaughter house, they're shackled upside-down and stunned or gassed unconscious, prior to having their throats slit.

CONCENTRATED ANIMAL FEEDING OPERATIONS

Frank Perdue's methods in creating the all-efficient animal factory were imitated far and wide in the poultry business. The practice, known as Concentrated Animal Feeding Operations (CAFOs), spread to pigs, cattle, and fish.[3] Massive use of antibiotics has allowed more and more animals to be crowded into less and less space. Once it was discovered that a continual low dosage of antimicrobials caused the animals to grow faster, their non-therapeutic use skyrocketed. CAFO animals have been transformed into living production machines through genetic engineering, selective breeding, growth hormones, and a steady diet of drugs.

By far and large, the people running these operations are chief executive officers of major corporations that have absolutely nothing to do with farming. This small group of corporate kings spend millions of dollars every year advertising how great their products are and what great deals they're giving us, while doing everything in their power to put real farmers out of business. They're doing this in one of two ways—by shutting them down outright or by turning them into contract farmers. Contract farmers put up all the money for the buildings and infrastructure, take all the risk, and are paid a pittance for their product—just enough to keep the wheels turning down on the farm. The handful of corporations controlling this major industry have been calling the shots (along with a few others at the top of this food pyramid) in government policy involving food production for decades. In short, this is an incredibly powerful industry that the U.S. Environmental Protection Agency (EPA) must somehow try to regulate.[4]

According to EPA estimates, there are now about twenty thousand CAFOs in the United States.[5] Of course, one has to be careful in painting all of these CAFOs with too broad a brush stroke. When properly managed, located, and monitored, CAFOs can provide an environmentally sound and humane source of meat, eggs, and milk. Nonetheless, CAFOs present major environmental problems.

The main environmental problems caused by CAFOs can be summed up with three words: *too much manure*. When you have thousands of animals packed together where the only thing to do is eat, that adds up to a lot of manure. Pigs can generate four times more excrement than humans, and cows generate fifteen to forty times as much. Before CAFOs, farmers spread barn manure and stall bedding on their fields. They still do on traditional farms. Besides being a source of free fertilizer, it was also an excellent soil conditioner, resulting in a sustainable practice where the animals fed the fields and the fields fed the animals. This doesn't work with CAFOs because there's too much manure and not enough field acreage to spread it on. With this option off the table, CAFO operators dump it into huge unlined lagoons the size of a football field (or larger) and somewhere around eight feet deep (or deeper). However, *lagoon* in this context is highly misleading because it conjures up images of a tropical paradise. The CAFO version of a lagoon is a stinking lake of feces, urine, and water. This is only a temporary solution because these lagoons soon fill up.

One solution is to follow the traditional path and give the manure to local farmers as free fertilizer. But this option is limited because many farmers don't want CAFO manure anywhere near their fields. This is such a problem that the CAFO industry has an acronym for it: WTAM—*willing to accept manure*. Farmers are not keen on applying manure loaded with antibiotics and hormones and all the other additives CAFO animals ingest, because they care about the long-term health of the soil. They also avoid it because it stinks to high heaven. This "stink factor" is a serious problem for the poor (and often primarily black) people living near CAFOs. Besides destroying any desire to go outside, the odor works its way indoors and permeates people's clothes. You simply can't get away from it. Some people have begun to fight back, winning large damages through public nuisance lawsuits.[6] While we focus on the water quality impacts in this chapter, air quality and greenhouse gas emissions are also major CAFO issues.

With many local farmers unwilling to accept CAFO manure, the contract producers are forced to transport it over long distances so they can distribute it over enough farmland to comply with Clean Water Act standards. But there's another problem that has grown in proportion to the growth of CAFOs: Specific animal-type CAFOs are now concentrated in distinct geographic regions. Hog production is concentrated in Iowa and North Carolina. Most beef cattle feedlots are located in five western Great Plains states. Broiler chicken production is concentrated in the Delmarva Peninsula and the Southeast. The upshot of all this regional CAFO expansion is that it's becoming even harder to find places to get rid of the ever-increasing amounts of manure in a way that avoids the EPA fining them or taking them to court. On top of these substantial problems, all that watered-down lagoon manure is heavy, making it very expensive to transport long distances.[7]

An alternative is for contract producers to spray it on their own fields, although these have been whittled down to make room for all the CAFO-related buildings. Nonetheless, this option is being utilized to the max. Under EPA regulations, they're not supposed to spray when it's raining. This is a big problem in places like North Carolina where it rains a lot.

CAFO operations are typically located near a body of water. During heavy rains, the fields can become manure-laced rivulets that empty into the nearest ditch, which in turn empties into the nearest water body. CAFO manure also percolates into the soil, contributing to groundwater contamination. And once again, it's the rural poor who are suffering the primary consequences, because they're dependent on a well for their water supply. Even after the water becomes so polluted that they no longer drink it, many still use it for bathing, washing clothes, and even cooking.

CAFO manure can attain disaster proportions during a hurricane or tropical storm. When a hurricane strafes the region, hog lagoons flood waste into creeks and rivers. Massive contamination occurred when Hurricane Floyd hit the North Carolina coast in 1999. Fifty-five lagoons were overtopped by floodwaters and six were breached. During the record-shattering Hurricane Florence in 2018, about three dozen lagoons overtopped or suffered structural damage. A post-Floyd program to buy out farms in the floodplain helped reduce the damage, but some two hundred or so hog farms remain on flood-prone areas.[8]

The worst-case scenario is a lagoon dike giving way. The one that brought CAFOs to national attention occurred in June 1995. The dike around an eight-acre hog lagoon in North Carolina gave way and twenty-five million gallons of hog-shit soup was suddenly on the loose, flooding yards, fields, and waterbodies as the entire cesspool emptied out. A twenty-two-mile stretch of the New River was affected, resulting in massive fish kills, algal blooms, and fecal bacteria contamination of public water supplies. Contaminated marinas and recreational facilities were closed. Work days were lost due to illness. Downstream shellfish beds were closed. The local food web suffered longer-term effects.[9]

This hog lagoon disaster fired up environmental groups, who began demanding that the EPA get serious about regulating CAFOs. The Clean Water Act had identified CAFOs as "point sources," making any discharge from them illegal (even if unplanned or accidental), unless authorized by the terms of a permit. However, for the first couple decades after the Clean Water Act was passed, the EPA was too busy to deal with the CAFOs that were popping up here and there. The agency had its hands full dealing with the extensive problems that had been the motivational force behind the act—industry dumping unfettered amounts of chemicals into the nearest stream or river, and cities dumping vast amounts of barely treated (or untreated) human sewage. This was a time for CAFOs what the Wild West had been for outlaws.

In 2003, thirty years after the Clean Water Act first brought CAFOs into the regulatory fold, the EPA strengthened their regulations. In a particularly controversial move, all CAFO operators were required to apply for a discharge permit, regardless of any *intent* to discharge. The EPA's rationale for this "duty-to-apply" provision was based on the presumption that most CAFOs have the potential to discharge pollutants into waters of the United States. The agricultural industry countered that nowhere does the Clean Water Act talk about a "potential" to discharge. The 2003 rule also required CAFO operators to develop and implement nutrient management plans, if they applied waste on fields. Such plans had to include a buffer zone between the edge of their fields and any waterway. This was also highly controversial. Environmentalists argued that these vegetation buffers were often inadequate because nutrients (especially nitrogen) can soak into the soil and pass under them. CAFO operators balked at the loss of field space for spraying manure.

With neither agricultural interests nor environmentalists happy with the new rules, they both took the EPA to court. After considerable give and take, the "duty to apply" was vacated by the court. The EPA issued its final revised rules in 2011.[10] Under the new rules, only CAFOs that "discharge or propose to discharge" must seek a permit. Large CAFOs were still required to get a permit, but many smaller ones would be allowed to decide if they needed to apply for a permit. As a result, only about a third of CAFOs have pollution permits.[11]

The logical question was, what would happen if a CAFO discharges without a permit? In that event, the EPA ruled that the CAFO could be fined for the discharge as well as for failing to get a permit. CAFO operators cried foul. The EPA resolved this double jeopardy by allowing a CAFO to certify that it is "designed, constructed, operated, and maintained not to discharge." If the CAFO later discharges by accident, it still gets fined for the discharge but not for failing to get a permit. The EPA requires that wastewater containment (the lagoon) be designed for a twenty-five-year, twenty-four-hour rainfall event. Discharges resulting from a larger event are considered agricultural storm water and allowed by permit.

On top of the large number and increasing size of CAFOs, enforcement has taken a big hit. This didn't start with Trump. In 2016, the number of EPA inspections was less than half the 2009–2012 average.[12] Riverkeeper, part of the global Waterkeeper Alliance, is taking up some of the slack. Founded in 1999, Riverkeeper works to eliminate the impacts of CAFOs on waterways and to help enforce environmental laws by publicizing violations and violators. Members of the group are on the move, binoculars handy, as they patrol back roads documenting waste violations. CAFO operators, along with their families and neighbors, are equally vigilant keeping an eye out for Riverkeeper vehicles. Confrontations are frequent, aggressive, and occasionally violent.

In the long run, cooperative approaches are likely to be more effective. A good example involves California dairy farmers. (California is the leading dairy state.) In the winter of 1996/1997, a series of subtropical storms brought more than thirty inches of rain to already saturated soils. The storms flooded many California dairies. Recognizing a need to be proactive, a group of dairy producers, government officials, and university specialists created the California Dairy Quality Assurance Program (CDQAP). In 1998, the CDQAP began delivering science-based workshops on food safety, animal welfare, and environmental stewardship, along with practical guidance for California dairy producers to help meet regulations. The EPA joined in 1999 and committed over four hundred thousand dollars in grant funding. Nearly eight hundred California dairy farms are now CDQAP certified through voluntary on-farm evaluation. This award-winning program celebrated its twenty-year anniversary in 2018.[13]

THE CLEAN WATER ACT

As hard as it is to believe, prior to the 1970s much of the sewage in this country was dumped into rivers with little or no treatment. It also was considered perfectly acceptable behavior to dump toxic industrial wastes into the closest stream or river. This was just plain old everyday normal behavior, like blowing cigarette smoke into someone's face. With the public clamoring for action, the Clean Water Act set a national goal to achieve *zero discharge* of pollutants into the nation's navigable waters—in basically ten years. This completely unrealistic goal was in line with a growing conviction that no one has a right to pollute the public's waterways. When it came time to pin down the actual rules of the act, Congress was much more realistic—anyone has a right to discharge pollutants into the nation's waters, as long as they obtain a permit and obey its requirements.

Despite having created the EPA two years earlier by executive order, President Nixon was opposed to the Clean Water Act. He was particularly troubled by the fourteen billion dollar price tag (in 1972 dollars) for federal grants to build or upgrade sewage treatment plants.[14] Senator Edmund Muskie, one of the authors of the act, warned that without these upgrades the nation's rivers would continue to serve "as little more than sewers to the seas."[15] Nixon remained unconvinced. When the water act finally reached his desk, he vetoed the bill, declaring, "Even if the Congress defaults its obligations to the taxpayers, I shall not default mine." Congress overrode the veto and the Clean Water Act was enacted into law on October 18, 1972.[16]

Before the 1972 water act, states set their own standards for what could be dumped into rivers. If they made things too difficult you could just move your operation to a more business friendly state, of which there were no lack. Then along came the Clean Water Act and suddenly it didn't matter if you were in West Virginia or Oregon, you had to meet the same pollution control standards—or pay a fine and then meet them.

Over the next decade, the EPA identified the best "practicable" pollution control technology for different industrial categories. To obtain a permit, an industry didn't have to use the specific technology identified by the EPA, but it had to do *at least as well* at pollution control. This turned out to be a smart move, because it encouraged innovations in better and more cost-effective technology. The EPA initially focused on reducing conventional contaminants, such as bacteria and oxygen-depleting substances that had been killing fish and harming aquatic life. Incrementally, more stringent control technology broadened pollution control to many toxic pollutants.

The EPA addressed the sewage problem through the largest federal public works program in human history.[17] From 1972 to 1990, federal grants totaling fifty-three billion dollars helped support more than eighty billion dollars in municipal wastewater treatment projects.[18] In 1987, the grants program was replaced with a federal revolving loan fund that provides low-interest loans to municipalities.

Through these efforts, the EPA greatly reduced sewage and industrial point source discharges to the nation's waterways. The real test of pollution control, however, is the quality of water in rivers, lakes, and streams. To address this issue, states (with EPA oversight) monitor each major waterbody and report regularly on those not meeting standards based on their designated uses (drinking water, fish habitat, swimming, and so forth). This turns out to be quite a surprising number. Currently, on a national level, over half of the assessed rivers and streams are not meeting water quality standards, as well as 70 percent of assessed lakes. If you added up all the miles of rivers in the United States that are "impaired" (polluted above water quality standards), it would stretch to the moon and back. The area of lakes with serious pollution problems encompasses an area larger than Switzerland.[19] Nonpoint sources are the primary culprit. Agriculture, and to a lesser extent urban storm runoff, are now the leading causes of water pollution.

The 1972 Clean Water Act failed to address nonpoint-source pollution, which means that the EPA has virtually no regulatory power to deal with it. As a result, the EPA's primary role to control agricultural and urban storm runoff pollution is to provide grants to support state and local efforts. Cities have more resources and concentrated authority, making urban runoff easier to address. Under the 1987 amendments to the Clean Water Act, the EPA began to issue permits for control of urban storm water that are based on best practices for reducing this nonpoint source of pollution. These practices often focus on green infrastructure, such as permeable pavement and filtering storm runoff through vegetation and wetlands before discharging to streams.

Storm runoff is a particularly serious problem in many old cities that have combined sewers, where wastewater from homes and businesses flows through the same pipes as storm water. During storm events, the system's treatment capacity can become overwhelmed and a noxious mix of untreated sewage and storm runoff overflows into waterbodies. To deal with this problem, some cities have constructed mammoth underground tunnel systems to store the excess water. After the storm passes, the water is sent to the treatment plant. The costs are immense, yet these underground tunnel systems combined with green infrastructure have substantially reduced urban runoff pollution. If a city is willing to swallow the price tag and rise to the engineering challenges, the bottom line is that urban storm runoff has at least a partial solution. Agriculture does not.

AGRICULTURAL POLLUTION AS A WICKED PROBLEM

The expansion and intensification of agriculture over the past decades have resulted in impressive achievements in food production and security, but this has come with a huge price tag to the environment. Covering about 40 percent of all land in the lower forty-eight states, agriculture is the major source of nutrients, pesticides, and sediment to America's streams, lakes, and estuaries.[20] Increased levels of nutrients stimulate algal blooms that harm the ecology of streams and lakes. Nitrate and pesticides move through the soil to groundwater, where they often show up in drinking water supplies. Almost everyone living in rural areas gets their drinking water from wells, many of which are contaminated by agricultural chemicals. Elevated nutrients in surface water increase water treatment costs for drinking water. Chemicals associated with agricultural activities empty into estuaries, where they harm valuable commercial and recreational fisheries and stimulate harmful algal blooms along the nation's coasts.

Agricultural pollution is a classic example of a "wicked problem"—one that has no definite solution *and* serious disagreement over the nature of the problem itself. With a large number of stakeholders and opinions involved, most wicked problems are connected to, or at least symptomatic of, other problems. Thus, the bottom line: A wicked problem can get better or it can get worse, but it never can be solved. The goal is to get into the "better" range, which requires practical approaches that are agreed upon by many diverse stakeholders.[21] The complexity of addressing agricultural pollution is well illustrated by the states draining to the Gulf of Mexico.

The dead zone that forms every summer in the Gulf of Mexico is the largest in the United States and the second largest in the world, surpassed only by a vast stretch in the Baltic Sea. This huge expanse of oxygen-depleted water (hypoxia) begins to form when spring floodwaters, rich in nitrogen and phosphorus, flow into the Gulf from the Mississippi and Atchafalaya Rivers. The nutrients from draining America's agricultural heartland feed immense blooms of algae that soon die and sink to the bottom, soaking up the available dissolved oxygen as they decompose. The main culprit is agricultural fertilizer and manure, which contribute over 70 percent of the nutrients that are plaguing the Gulf.[22] The resulting dead zone often grows larger than five thousand square miles and stays in place until autumn, when a tropical storm or frontal weather systems cause the surface and bottom waters to mix.

In 2001, a joint federal–state Hypoxia Task Force led by the EPA set a goal to reduce the areal extent of the Gulf's dead zone by about two-thirds. They gave themselves until 2015 to do it. Despite these good intentions, the size of the hypoxic zone has remained essentially the same. The goal has now been extended to 2035, but to achieve it, nitrogen loads must be cut by more than half.[23] Further complicating this granddaddy of wicked problems, improving fertilizer management is not enough. To have even a hope of achieving this goal will require major efforts to capture nutrients through drainage control, restoring wetlands, and planting buffer zones along streams and rivers.[24]

In 2017, the dead zone was the largest ever recorded—at more than eighty-seven hundred square miles, about the size of New Jersey. This blockbuster was caused by an unusually wet May that resulted in the equivalent of twenty-eight hundred train cars of fertilizer being dumped into the Gulf.[25] At the time when the Clean Water Act was passed, the poster child for the country's water problems was a burning river. Today, it's a dead zone in the Gulf of Mexico that staggers the imagination.

The Mississippi River has the world's fourth-largest drainage basin, encompassing all or parts of thirty-one U.S. States and two Canadian provinces. The number of farms is legion. Almost every one of them, to one degree or another, is flushing nutrients into the Gulf of Mexico. It's a tragedy of the commons on a mega-scale. The EPA is working to reduce the size of the hypoxic zone principally by encouraging states to develop and implement nutrient reduction strategies. The EPA's role is all carrot. It's up to the states to apply any stick.

Minnesota, where the Mississippi River begins at Lake Itasca, has been among the most aggressive states in taking up this challenge. The state takes pride in being the Land of 10,000 Lakes (the number is actually closer to twelve thousand), and they care about water quality. As one of the more environmentally progressive states, Minnesotans have taken the virtually unheard-of step of taxing themselves for clean water. In 2008, voters approved a three-eighths of one cent increase in the state's sales tax dedicated to improving the environment and culture. One-third (about one hundred million dollars) is earmarked for clean water. [26]

Minnesota's ambitious plans to reduce nutrient loads have not been without controversy. For example, in 2015, Minnesota Governor Mark Dayton signed the Buffer Strip Law, requiring grassy buffers averaging fifty feet wide along streams and a minimum of 16.5 feet wide along public ditches to protect the waters from soil erosion and fertilizer runoff. Cost-share funds are available through the state to help pay for the buffer strips. Many farmers protest the law because it takes valuable land out of production. They see it as an unconstitutional land grab. [27]

Voluntary efforts have fared better. In 2012, the state, the EPA, and the Natural Resources Conservation Service teamed up to create the Minnesota Agricultural Water Quality Certification Program. Farmers who join the program receive priority for technical assistance and cost-share dollars. A major enticement is that certified farms are automatically assumed to comply with any new water quality rules for ten years. They also receive public recognition, along with a sign to display at the entrance to their farm.

Certification is a two-step process. A computerized assessment evaluates factors such as fertilizer application and tillage practices. Then the certifier and farmer walk the farm field by field, discussing conservation practices that could be utilized. Jared Nordick, a farmer in northwest Minnesota, jumped at the chance to have his farm certified first and is considered a star in the program. His father, Jerry Nordick, was initially a skeptic, but now sees the advantages: "I've said for years, there are many farmers doing a good job, but don't get the recognition. This is an opportunity for us to communicate our water-quality efforts."[28]

It's not just streams and lakes that have high nitrate levels, public and private well owners are grappling with the same problem. In recent years, the state has drafted a Groundwater Protection Rule with substantial input from farmers and other stakeholders, including seventeen public meetings held across the state. The final rule, which was scaled back from earlier versions, restricts fall and winter applications of nitrogen fertilizer in about 12 percent of state cropland and establishes best practices immediately around public supply wells having the highest nitrate levels. The latter effort is voluntarily based on recommendations from local advisory teams, but regulations come to bear if not enough progress is made. Leading farm and environmental groups critical of early versions of the proposal say they can live with the revised plan. The president of the Minnesota Corn Growers Association called it "a reasonable approach." Nonetheless, Republican legislators (unsuccessfully) tried to kill the rule.[29]

Overall, Minnesota's efforts to address surface water and groundwater contamination are unlikely to attain their ambitious goals, but the state is making an impressive concerted effort.

THE CHESAPEAKE BAY

The Chesapeake Bay, the nation's largest estuary, is an example where incremental progress is slowly being made with oversight by the EPA. More than eighteen million people live in the Chesapeake Bay watershed, which starts as far north as New York and encompasses six states and the District of Columbia. Renowned for its blue crabs, oysters, and striped bass, the bay is a magnet for sport fishermen, boaters, and tourists from far and wide. This is a place where you're transported back to a simpler time. The bay's 11,500 miles of shoreline can challenge the destination oriented. You need explicit directions (and some luck) to find such places as the Riverside Inn outside of Annapolis, where tablecloths are huge sheets of butcher paper that are covered with heaped baskets of crabs and oysters fresh off the boat. After washing it all down with cold steins of beer (there's soda for the kids), you can stroll down the pier and watch the fishing boats pull in and unload. The bay's weather-scarred watermen aren't putting on a show. This is their livelihood, which involves the strenuous discipline of kicking yourself out of bed when it's still dark so you can be on the water at first light, heading out into whatever the weather happens to be that day—rain, sleet, snow, wind, and often some combination. God love those summer days when you can work your trap lines under a blue sky, the sun warming your back.

Ducks, geese, and great blue heron are just a sampling of the many bird species that live here or drop in for a few days to rest and refuel at this five-star stopover on the Atlantic Flyway. On the surface it all looks great, but the watermen will be the first to tell you that all is not well in the Chesapeake Bay. Underneath that shimmering, placid water is the equivalent of depressed neighborhoods and virtual ghost towns. For decades, the bay has been imperiled by nutrient overloading from agriculture, urban runoff, wastewater discharges, and air pollution. There's a lot of blame to spread around, but everyone knows it's mainly agriculture that's endangering the bay.

Excess nitrogen and phosphorus have created a domino effect. The nutrients stimulate algal blooms, which decompose and cause large areas of low dissolved-oxygen concentration that kills aquatic life. Crabs and other slow-moving, bottom-dwelling organisms are particularly vulnerable. Algal blooms also block sunlight needed by submerged grasses. When those grasses die, they remove an important food for waterfowl and shelter for crabs and young fish.

In 1983, with the Chesapeake reaching a critical tipping point, the governors of Maryland, Virginia, and Pennsylvania, the mayor of the District of Columbia, and the EPA administrator signed an agreement launching the Chesapeake Bay Program. After decades of worsening conditions, officials were finally taking action. Over the next couple decades two more agreements were signed, which basically boiled down to ambitious, but non-binding, goals to reduce nutrient loads to the bay. The agreements failed miserably. As time passed, the collaborative partnership was widely considered dysfunctional. Thanks to the Clean Water Act and infusion of federal dollars there was some progress on sewage treatment plants, but little was done to tackle nonpoint-source pollution from farms, septic tanks, and city storm sewers. The bay failed to improve.[30]

A turning point came in 2007 when the EPA and the bay jurisdictions (New York, Pennsylvania, West Virginia, Delaware, Maryland, Virginia, and the District of Columbia) agreed to establish a multi-state pollution diet, officially known as a total maximum daily load (TMDL). Previously having had a fairly passive role, the EPA was given primary responsibility for completion of the Bay TMDL, which involved estimating 276 nitrogen, phosphorus, and sediment TMDLs for ninety-two individual Chesapeake Bay tidal segments.[31]

In 2009, President Obama signed an Executive Order that declared the Chesapeake Bay a national treasure and directed multiple federal agencies to lead a renewed effort to restore and protect the bay and its watershed. The EPA worked closely with the bay states, as well as held many public meetings with farmers, developers and homebuilders, municipal authorities, local elected officials, and environmental groups. On December 29, 2010 (two days before a court-ordered deadline), the EPA finally established the Chesapeake

Bay TMDL. This monumental undertaking to restore the nation's most productive estuary calls for 20 to 25 percent reductions in nitrogen, phosphorus, and sediment. All pollution control measures need to be in place by 2025. Each state is responsible for meeting their part of the pollution diet. Milestones are set every two years to demonstrate progress. And if a state consistently fails to do its part, the EPA can take charge.

Since the EPA established the Chesapeake Bay TMDL in 2010, nutrient and sediment loads have fallen in places, and the bay has started to rebound. According to the Chesapeake Bay Foundation's 2016 annual report: "We are seeing the clearest water in decades, regrowth of acres of lush underwater grass beds, and the comeback of the Chesapeake's native oysters, which were nearly eradicated by disease, pollution, and overfishing." Even more progress was reported for that iconic symbol of the bay, the blue crab.[32] More good news came in 2018 when the University of Maryland Center for Environmental Science reported that the bay showed improvement in every region for the first time in the thirty-three years that scientists have assessed its health. "It seems that the restoration efforts are beginning to take hold," said Bill Dennison, vice president for science application at the university.[33] A separate study by a large group of researchers showed an unprecedented resurgence in aquatic grasses in the bay.[34]

This is all great news, but the bigger picture isn't so rosy. Overall, the bay is still getting a poor grade, particularly with the nitrogen goals. Nitrogen drives algae growth in the bay during most of the year. Future progress is in doubt because much of the gains to date came from upgrades of sewage treatment plants. And then there's the wild card—an enormous buildup of sediment behind the Conowingo Dam on the Susquehanna River at Maryland's border with Pennsylvania. This reservoir has almost filled with sediment. When it tops over, projected within the next few years, it will become useless. U.S. Geological Survey scientists estimate that when the reservoir reaches capacity, sediment and phosphorus from the Susquehanna River would increase by about 250 percent and 70 percent, respectively.[35]

Voluntary, collaborative measures have been applied longer in the Chesapeake Bay than in any other ecosystem-wide restoration program in the world, with billions of federal cost-share dollars invested.[36] Yet progress has been limited because animal wastes and chemical fertilizers contribute the largest source of pollutants. More than eighty-three thousand farms make up the ten billion dollar agricultural industry in the Chesapeake Bay watershed.[37] And then there's the massive amounts of storm water runoff from urban and suburban areas.

The politics are daunting. The bay restoration effort has met with political resistance from every direction—congressional, state, and local. As Oklahoma's attorney general, Scott Pruitt participated in a legal challenge to the bay restoration program by farmers, homebuilders and other stakeholders. House Republicans continue to try to add provisions to spending bills that would bar the EPA from enforcing the bay TMDLs.

A major reason why the bay's overall progress is moving along at a crab's pace is due to the six watershed states having varying levels of skin in the game.[38] The Susquehanna River, the leading contributor to the bay's nitrogen loads, is located almost entirely in Pennsylvania, with its headwaters in upstate New York. Fertilizer and manure nutrients picked up by the Susquehanna have a disproportionately large impact on the estuary's water quality, but Pennsylvania and New York do not reap the economic benefits of restoring the Chesapeake. Almost all the benefits go to the two states that border the bay—Maryland and Virginia.

Pennsylvania is known as the Keystone State because of its central location among the original thirteen colonies and its pivotal role in the Declaration of Independence and U.S. Constitution. Because of the Susquehanna River, the state is also the keystone to restoring the bay—and it's getting failing marks. By the end of 2017, Pennsylvania had achieved only 18 percent of its nitrogen reduction goal. If the state decides to get serious in meeting its 2025 goals, it will have to reduce 2.5 times as much nitrogen in less than eight years as it has in the past thirty-two years. If it fails to do so, much of the Chesapeake Bay will not attain its critical goals.[39]

Under pressure from the EPA, Pennsylvania has agreed to reboot the state's lagging efforts to meet the bay clean-up goals. Among other actions, improved manure management is viewed as the biggest challenge. By 2025, the strategy also calls for planting ninety-five thousand acres of riparian forest buffer along streams. This immense effort is spearheaded by the Chesapeake Bay Foundation, alongside many partners—national, regional, state, and local agencies; conservation organizations; watershed groups; businesses; and individuals. The strategy also commits to increasing the number of farms inspected each year, from less than 2 percent (that would take more than half a century to complete) to 10 percent. These inspections primarily focus on whether farmers have up-to-date plans required by the EPA for managing manure, erosion, and sediment.[40]

Much of the focus is on five counties in southcentral Pennsylvania. Jay Diller owns 350 acres in one of these priority counties, where he milks about 180 cows, raises young hens to be sold to egg operations, and grows crops. Diller is an avid reader of agricultural publications and is well versed in how to reduce soil erosion and responsibly manage manure. He practices no-till cropping methods and plants winter cover crops for environmental reasons, which he believes also makes good business sense. "The last thing I like to see is brown water crossing the road," Diller says. "I think about water quality, and I like to keep my soil on the farm." Nonetheless, Diller's erosion and sedimentation plan was out of date. "I find the paperwork part of all this frustrating," he says, "but if this is what it's going to take to improve, I say, let's do this. I'd rather do it than be in violation. I don't want to be in violation. Farmers get enough bad publicity." But Jay Diller is just one of the more than thirty-three thousand farmers in Pennsylvania's part of the Chesapeake Bay watershed, and not everyone shares his attitude.[41]

The Chesapeake Bay restoration program was decades in the making, is central to restoring the bay, and is beginning to show some hopeful signs of success. Yet, completely impervious to the progress being made and the EPA's essential leadership role, the Trump administration proposed a total elimination of the program in their first budget proposal. The reasoning went that, as a "regional effort," it should not be funded by Washington. Without pressure and oversight from the EPA, the bay will fail to meet its TMDL goals.[42]

Trump's first budget request also proposed gutting the EPA's Great Lakes Restoration Program, which provides funds to address the harmful algal blooms that are plaguing Toledo and western Lake Erie, while also fighting invasive species, removing contaminated sediments from harbors, and restoring wildlife habitat. Congress reinstated the funds for the two programs, but again the administration proposed gutting both programs by 90 percent in 2019 and 2020. Trump soon backed off the 2020 cuts to the Great Lakes funding for political reasons, but clearly has no interest in making either the Chesapeake Bay or the Great Lakes great again.

ADDRESSING A WICKED PROBLEM

Environmentalists often vilify farmers and vice versa, yet the reality is much more complicated. If we look at the entire continuum of traditional farmers to Big Ag food producers, on the whole they are serious about land stewardship. Nonetheless, it's a very touchy issue to tell farmers how to use their land—even when there's something in it for them.

A good starting point is soil health. Improved soils increase food production, and at the same time enhance water infiltration, filter and buffer pollutants, and support biodiversity. Steps to improve soil health through organic matter retention, erosion control, diversified rotations, and cover crops are a win-win for farmers *and* water quality. Yet getting farmers to adopt these approaches can be challenging. Cover crops are a case in point.

Driving across the rolling hills of Iowa after harvest, one sees corn stubble and open fields as far as the eye can see. What you don't see much of are winter rye and other cover crops that contribute needed organics to the soil and help keep nutrients in the ground for the next growing season. Sometimes called "catch crops" for their ability to retain nutrients, cover crops are a popular conversation topic across the midwest—but it hasn't gotten much beyond the talking stage. Kevin Ross, a sixth-generation family farmer and former president of the Iowa Corn Growers Association, encourages their adoption: "Like any new crop on the farm, cover crops have a learning curve," he says. "It takes some experimentation with varieties and seeding methods. It takes time to see benefits while identifying potential risks. The important thing is to get started!"[43]

There's a good reason to do so. Water quality in Iowa's rivers has deteriorated despite the state's nutrient reduction strategy.[44] The state's capital, Des Moines, operates one of the world's most expensive nitrate-removal facilities. In a high-profile case, Des Moines Water Works filed a federal lawsuit against drainage districts in three counties that are discharging high levels of nitrates into the Raccoon River, the source of drinking water for five hundred thousand central Iowa residents. The judge dismissed the lawsuit, saying Iowa's water quality problems are an issue for the Iowa legislature. The problem remains unresolved.[45]

The crisis in Toledo helped motivate Ohio to ban manure and fertilizer applications on frozen ground. Nonetheless, such commonsense practice is often fought by farmers because of the merest whiff of regulation. They also can't see nutrients running off their land in the same way they can watch sediment loss during a snowmelt or heavy rain. Nutrient management plans can be a tough sell and, even then, are not necessarily followed.[46]

Programs to get farmers to reduce environmental damage are most effective when delivered by people they trust. This is where the Natural Resources Conservation Service (NRCS), part of the Department of Agriculture, plays a key role. Everyone has heard of the EPA, but the NRCS remains under the radar outside the agricultural community. With its stated mission, "Helping People Help the Land," the NRCS provides essential assistance to address agriculturally related water quality problems. While the EPA wields both the carrot and the stick, the NRCS only dangles the carrot. It's been said that even John Birchers let the NRCS on their land.

The NRCS provides technical assistance and grants to encourage voluntary, incentives-based conservation approaches. These approaches often build on existing efforts, such as nonpoint-source grants from the EPA. Cost-sharing funds incentivize farmers to plant grassy buffer strips on the perimeter of fields to capture nutrients, restore wetlands on less productive land, improve manure storage, and fence off streams to keep out animals and their wastes. To get more bang for its buck, the NRCS focuses on targeted watershed projects for the most vulnerable lands. These volunteer efforts are not the complete answer, however, as reductions in nutrients, particularly nitrate, have proven elusive. As one NRCS employee puts it, "Both the carrot and stick are needed. I'm glad I'm the carrot."

A promising development in recent years is a national movement known as "One Water" that encourages agencies and others to work together to manage water more holistically.[47] It's a shift from the traditional separate management of drinking water, wastewater, and storm water. The focus is on collaborative and holistic approaches that achieve desired results most effectively and at lowest overall cost. The Yahara River watershed in southcentral Wisconsin is a good example of this approach.

Due to tightening regulations, the Madison Metropolitan Sewerage District needed to achieve a small increment of phosphorus removal that would cost $130 million for additions to its treatment plant. As an alternative to these expensive upgrades, the district spearheaded a partnership among cities, towns, farmers, environmental groups, the University of Wisconsin, and the Wisconsin Department of Natural Resources. Known as the Yahara Watershed Improvement Network, the partnership's projects include adding grass buffers between farm fields and waterways, sprinkling cover crop seeds from airplanes, manure management, and urban leaf collection. These are funded by the members, grants, and other sources. In a few years, the partnership had spent just seven million dollars and reduced the phosphorus load by thirty thousand pounds in 2016 alone. The goal is to achieve a yearly reduction of ninety-six thousand pounds by the project's end in 2036.[48]

As her term came to a close, EPA Administrator Gina McCarthy spoke about the agency's failure to build trust with rural America, particularly with farmers: "We work in a lot of communities; we work on a lot of urban-related issues. We have great relationships with mayors. But we have tended to not be able to have a very compelling rural agenda and to build constituencies there." Strengthening the collaborative relationship between the EPA and NRCS is a place to start. Private-public partnerships and the One Water movement are also helping, but the problem remains—how do you get farmers on board to substantially reduce nutrient loads to the environment. McCarthy recognized this dilemma, saying: "There has got to be some wake up calls, some aha moments, where agriculture demands [environmental change] of themselves, because I don't think we are in a position to demand it of them."[49]

5

INCONVENIENT CONNECTIONS

We cannot expect to preserve the remaining qualities of our water resources without providing appropriate protection for the entire resource.

—Senator Howard Baker (R-TN), 1972[1]

Over 40 percent of the water we drink and almost half of irrigation water used to grow our food comes from groundwater.[2] But this critical resource is out of sight and (for most people) out of mind, making groundwater the neglected child of the water world. This problem is far from new. In 1861, an Ohio court famously concluded that groundwater was so "secret, occult, and concealed" that any attempt to regulate it "would be involved in hopeless uncertainty, and would be, therefore, practically impossible." The state clung to this view for more than one hundred years.[3]

Groundwater and surface water are joined at the hip. During dry periods, groundwater sustains streams, lakes, and wetlands. The connection goes both ways as surface water donates generously to groundwater. This ongoing give and take provides a natural pathway for contaminants to move up and down from one resource to the other. While the need to protect what's down-under is obvious, the Clean Water Act fails to recognize these connections.

The act even takes it a step further, clearly stating that the U.S. Environmental Protection Agency (EPA) does not have authority to regulate groundwater. Even so, questions remain that are not so clear. For example, what about a point source that's directly polluting groundwater, and then those contaminants show up in surface water? As one court noted, "it would hardly make sense for the Clean Water Act to encompass a polluter who discharges pollutants via a pipe running from the factory directly to the riverbank, but not a polluter who dumps the same pollutants into a man-made settling basin some distance short of the river and then allows the pollutants to seep into the river via the groundwater."[4]

Some courts have recognized that the Clean Water Act requires permits for discharges to groundwater when there is a "direct hydrological connection" to surface water.[5] At face value, this *conduit theory* is hard to argue with. However, it's vague on key factors such as the time it takes for a pollutant to move to surface waters, the distance it travels, and its traceability to the point source. Furthermore, states are responsible for the groundwater within their state and look very unfavorably at any action they feel is federal overreach.

In one of the surprises of Supreme Court cases, ultraconservative Justice Antonin Scalia supported the conduit theory in the 2006 *Rapanos v. United States* case. Justice Scalia observed that the Clean Water Act broadly regulates pollutants from point sources *to* navigable waters, but nowhere does it say they have to discharge *directly to* navigable waters. He concluded, "even if pollutants discharged from a point source do not emit 'directly into' covered waters, but pass 'through conveyances' in between," they would still violate the act.[6] Court cases continue to test this idea.

The Lahaina Wastewater Reclamation Facility, on the Hawaiian island of Maui, pumps several million gallons of treated sewage a day down four wells. From there, the water flows through the porous lava rock into the ocean, a half-mile away. There is no doubt that the wastewater has a direct hydrological connection to the ocean via groundwater. A dye study funded by the EPA and the Army Corps of Engineers (hereafter Corps) demonstrated that water from two of the injection wells reached the ocean in less than three months. The enriched nitrogen and phosphorus in the effluent is causing ecological harm to a once-pristine coral reef and popular recreational spot. If instead of using the wells, the treatment plant discharged its wastewater in a pipe directly to the ocean, it would require a permit under the Clean Water Act.[7]

Under the citizen suit provision of the Clean Water Act, the Hawaii Wildlife Fund and other environmental groups sued, asserting that the wastewater agency must obtain a permit for these "point-source" discharges. In February 2018, the Ninth Circuit Court of Appeals agreed, calling the well injection the "functional equivalent" of a direct discharge into navigable waters. "At bottom, this case is about preventing the county from doing indirectly that which it cannot do directly," the court concluded.[8]

A few weeks prior to the ruling on the Maui case, a district court had rejected a similar lawsuit brought against the E. W. Brown power plant in Kentucky. Selenium and other trace elements were leaching from coal ash ponds into groundwater, which carried the contaminants into a nearby lake where they were causing physical deformities in fish.[9] The direct hydrologic connection was well documented, but in this case the court ruled that the Clean Water Act did not apply.[10] This debate is just one part of broad-based controversies over coal ash ponds.

COAL ASH

Coal ash is the residue left after coal is burned to produce electricity. It is stored in pits, often mixed with water. Most coal ash pits are unlined and located alongside waterways. The scope of this problem is immense. At its peak, enough coal ash was created each year to fill railroad cars stretching between the North and South Poles. Only household trash exceeds coal ash in the volume of solid waste generated. Like coal itself, the ash contains toxic elements such as arsenic, mercury, and selenium that contaminate groundwater, streams, and lakes. The problem is also a case of environmental injustice. Seventy percent of coal ash dumps across the country are located in low-income communities, including Native American tribes.

The Moapa Band of Paiute Indians' reservation lies about fifty miles northeast of Las Vegas. Immediately next door and directly upwind lies the now-shuttered Reid Gardner coal-fired power plant. For decades, during the all-too-common windy days in this desert community, tribal members became imprisoned in their own homes. Going outside was like walking into a dust storm loaded with arsenic, lead, and other toxic elements. Tribal members experienced high rates of asthma and heart disease, as well as general fatigue and frequent headaches. When the fight against coal ash began around 2000, William Anderson, the tribe's chairman, noted that a majority of children now had asthma or a breathing illness, but "no one would listen to what we were saying."[11]

Anderson became a passionate activist, leading a three-day, fifty-mile march from the Reid Gardner coal plant to downtown Las Vegas.[12] When Las Vegas residents became aware that some of their power was coming at the expense of the tribe's health, many joined the campaign to close the plant. Working with the Sierra Club and Earthjustice, the tribe filed a lawsuit against NV Energy, the plant owner. The company eventually settled for $4.3 million in 2015. The plant closed in 2017. The settlement money was used for a community health center and to help the tribe buy water rights, monitor air quality, and oversee cleanup of pollution along the Muddy River, which runs through the reservation. But the coal ash remains a threat to air and groundwater.

Under Anderson's leadership, the tribe completed a deal for a solar farm on the reservation as an alternative source of local jobs. This facility generates electricity for more than a hundred thousand homes in Los Angeles, contributing toward California's ambitious renewable energy goals. "We showed them that there's an effective way to have a solution instead of destroying the environment, plants and animals and spoiling the water," Anderson commented.[13] William Anderson died in 2018, at the age of forty-four. During the last years of his life, he was in and out of the hospital suffering from multiple illnesses and mysterious infections.

For decades, coal ash stored at more than four hundred coal-fired power plants across the country remained one of those unrecognized problems in plain sight. At one o'clock in the morning on December 22, 2008, it suddenly jettisoned to national attention when a coal ash dike failed at a coal plant in Kingston, Tennessee. Over five million cubic yards of coal ash slurry flooded more than three hundred acres, dumped waste into two nearby rivers, damaged numerous homes, and made several homes uninhabitable. Cleanup took six years and over a billion dollars.[14] In response, the EPA inspected the structural integrity of over five hundred coal ash impoundments across the country. More than one-quarter were rated as "poor."[15]

Another shoe fell in 2014, this time in North Carolina, when a security guard noticed low water levels in a coal ash pond at Duke Energy's Dan River site. A broken drainage pipe beneath the pond was emptying coal ash directly into the Dan River, with tens of thousands of tons of coal ash contaminating more than seventy river miles. Having failed to repair corroded pipes at this site and illegally discharging coal ash into waterways at other sites, the utility settled criminal charges for $102 million. The plea agreement followed a ninety-minute court session in which a Duke Energy lawyer repeated the words "guilty, your honor" more than twenty times.[16]

Air pollution at the Paiute reservation and catastrophic failures in Tennessee and North Carolina are easily recognizable risks from coal ash pits. Groundwater contamination is much subtler, yet the sequence of events is predictable. A coal-fired power plant is located next to a river or lake to provide cooling water for the plant. The residual coal ash is dumped into unlined pits adjacent to the waterbody. These ponds leak directly into shallow groundwater that is connected to the river or lake. In other words, coal ash ponds are just another version of point source pollution.

The Vermilion coal-fired power plant in Illinois is a textbook case. Crossing central Illinois from the west, the "flat-as-a-pancake" farmland topography abruptly changes upon approaching the Middle Fork of the Vermilion River. The clear, fast-moving water has carved a meandering path through glacial deposits, punctuated with pools and riffles. It's a popular spot for kayaking, tubing, or swimming on hot summer days. Sycamores line the riverbanks, with oaks, beeches, sugar maples, and dogwoods populating the uplands. Wildlife is abundant. [17]

In a seventeen-mile section of river designated as the only National Wild and Scenic River in Illinois, the otherwise bucolic scenery is marred by the now-defunct Vermilion coal-fired power plant, located just downstream from a popular canoe and kayak launch. From the mid-1950s until 2011, the plant generated 3.3 million tons of coal ash that was dumped into three massive unlined pits in the Middle Fork floodplain. The plant owner (Texas-based Vistra Energy) and the state of Illinois acknowledge that groundwater beneath the pits is contaminated with arsenic, chromium, and other trace elements that discharge to the Middle Fork. Seeps from groundwater stain the riverbank an unnatural orange-red color, and rust-colored water pools along the river's edge. Adding to the concerns, the meandering river is eroding the river bank next to the pits, raising worries of a catastrophic spill similar to the disasters in Tennessee and North Carolina.

In 2018, the nonprofit group American Rivers named the Middle Fork as one of America's ten most endangered rivers, citing the threats to recreation and aquatic life. Environmentalists want the coal ash removed to a safe and properly designed facility away from the river. Vistra's solution is to cap the coal pits and leave the ash permanently in place behind a wall of giant rocks along a stretch of riverbank six football fields long. In 2018, citing the Clean Water Act, environmental groups sued Vistra over the toxic metals leaching from the coal ash pits into groundwater and from there to the Middle Fork. A federal judge dismissed the case. [18]

In 2015, the Obama administration rolled out the first federal regulations for disposing of coal ash from power plants. Prior to this time, it had been virtually unregulated. The rule, established under the Resource Conservation and Recovery Act, set structural integrity standards for disposal sites and required new coal ash ponds to be lined. The rule also required utilities to begin testing the groundwater near coal ash sites and making the data public. Coal ash ponds found to be contaminating groundwater, or seismically unsafe, would be closed.

The new law was developed after many years of negotiations with utilities, other affected industries, and environmentalists. It also contained several compromises. The law didn't apply to sites closed prior to 2015, such as the Vermilion power plant. (This provision was overturned in 2018 by the DC Circuit court.[19]) In addition, the EPA decided not to classify coal ash as hazardous waste. Environmentalists had fought long and hard for this classification, because it would trigger federal enforcement. Instead, the law is framed as "self-implementing," which means that if a company doesn't comply, the only redress is a lawsuit. EPA Administrator Gina McCarthy called the rule a "pragmatic step forward" to protect communities from coal ash.[20] At the end of the Obama administration, Congress gave states additional flexibility to adopt state-specific closure plans, as long as they are *as protective* as the federal rule.

Not classifying coal ash as hazardous waste makes a certain amount of sense. More than 40 percent of coal ash is recycled for making building materials, such as concrete and wallboard. In particular, the cement industry commonly mixes "fly ash" into cement. This makes cement green in color, and by substituting coal ash for some of the cement, it's "green" in another important way. Cement, the key ingredient in concrete, contributes up to 10 percent of global carbon dioxide emissions—surpassed only by transportation and energy. Using fly ash in cement is a double win—it reduces waste and the impact of concrete production on global warming. If the EPA had decided to designate coal ash as hazardous waste, the costs of handling the material would have skyrocketed.

Key industry players viewed Obama's coal ash rule as acceptable. Then along came Donald Trump's broad-based attack on coal regulations. With this opening, the Utility Solid Waste Activities Group, a trade association representing more than one hundred power companies, petitioned the EPA to weaken monitoring and remediation requirements.

In March 2018, just as initial data required by the coal ash rule were revealing groundwater contamination at numerous sites, the Trump administration proposed to "incorporate flexibilities" into the rule. The proposed revisions would extend the life of some existing coal ash pits for a year and a half. This extension seems reasonable for such a large problem, particularly if it was tied to a plan for more ash recycling. More problematically, the proposed revisions would allow state directors to shorten monitoring of sites that show no evidence of groundwater contamination, thereby ignoring the fact that groundwater moves slowly. Avner Vengosh, a Duke University expert on the environmental impacts of coal ash, argues, "We have very clear evidence that coal ash ponds are leaking into groundwater sources. The question is, has it reached areas where people use it for drinking water? We just don't know. That's the problem."[21] Without responsible monitoring requirements in place, officials are unlikely to know.

The Trump administration's proposed coal ash revisions would roll back an important rule affecting the nation's water resources. Yet Pruitt's EPA held only one public hearing on the proposal—in a hotel near Washington, DC, far away from most coal ash ponds. Seventy people from across the country traveled to the hearing to speak up about why coal ash disposal regulations should not be relaxed. Participants included environmentalists, people living next door to coal ash, a nurse, a pediatrician, college students, tribal members (including the Moapa), state legislators, a Girl Scout from Illinois, and a small-town mayor who received a rousing applause when he suggested renaming the EPA as UPA—Utilities Protection Agency.[22] The public's reasoning and pleading fell on deaf ears. Four months after proposing the revision (and a week after Pruitt resigned), Andrew Wheeler, a former coal industry lobbyist, signed his first rule as acting EPA administrator. The revisions to the coal ash rule under the Resource Conservation and Recovery Act were now official.[23]

The applicability of the Clean Water Act in situations where pollution from coal ash ponds or other point sources reaches waterways via groundwater (the conduit theory) remains unresolved. The

U.S. Supreme Court recently agreed to hear the case of the Lahaina wastewater facility in Hawaii. In the meantime, the EPA issued its own interpretation. Instead of trying to define what is meant by a "direct hydrologic connection" for purposes of the law, Trump's EPA asserts that once pollution travels through groundwater, it "breaks the causal chain" between a pollutant source and surface waters.[24] This, of course, is quite contrary to hydrologic reality.

WATERS OF THE UNITED STATES

An even larger controversy swirls around the basic question of which wetlands and other waterbodies are actually covered by the Clean Water Act. Drainage of wetlands has been going on since Europeans first settled Colonial America. Encouraged by federal laws that provided incentives for "reclaiming" swamps and marshes, more than half of the country's original wetland area has been lost. Several Midwestern states have lost more than 85 percent; California more than 95 percent.[25] These swamps and marshes were considered just nuisances standing in the way of land development and agricultural production, while also being a breeding ground for malaria and other diseases. Today, it is well known that wetlands provide invaluable habitat for waterfowl, fish, and other wildlife. About a third of North American bird species rely on wetlands for water, food, shelter, and breeding.[26] Less appreciated are the water quality benefits that these "kidneys on the landscape" provide in removing nutrients, pesticides, and toxic metals from waters flowing through them.

In 1989, President George H. W. Bush, an avid fisherman, announced a policy of "no net loss" to protect wetlands. "Wherever wetlands must give way to farming or development," Bush told a meeting of Ducks Unlimited, "they will be replaced or expanded elsewhere."[27] Although this has remained a national goal, the net destruction of wetlands has slowed, but not stopped.

The Clean Water Act is the primary federal law regulating wetlands. Section 404 of the act prohibits discharging dredged or fill material into "navigable waters" without a permit from the Corps. The intent of the Clean Water Act was to restore and protect more than just traditional navigable waters, by equating "navigable waters" with "waters of the United States." The problem is that the term "waters of the United States" leaves much room for interpretation. As Oliver Houck, an expert on water quality law, wrote in 1989, "Section 404 of the Clean Water Act lies like an open wound across the body of environmental law, one of the simplest statutes to describe and one of the most painful to apply."[28]

The role of the Corps can be traced back to the earliest federal law related to water quality, the 1899 Refuse Act, which required a permit from the Corps for discharge of any "refuse" into navigable rivers and harbors. The EPA shares enforcement responsibility with the Corps, while also providing guidance and oversight. For almost thirty years, the "waters of the United States" was broadly interpreted as most surface waters and wetlands, including those adjacent to tributaries of navigable waters, as well as "isolated" ponds and wetlands of ecological significance. In 2001 and 2006, two U.S. Supreme Court decisions considerably muddied the jurisdictional scope.

The first case, commonly known as *SWANCC* (named after the plaintiff, Solid Waste Agency of Northern Cook County), involved a proposed facility for disposal of domestic solid waste at abandoned quarries that provided habitat for migratory birds. The Corps claimed jurisdiction over this site based on its self-declared Migratory Bird Rule, which asserted jurisdiction over ponds and wetlands that provide habitat for migrating waterfowl. The high court disagreed, ruling that the Migratory Bird Rule exceeded the Corps' authority. This decision created uncertainty about the Corps' regulatory jurisdiction over any "isolated" water body.[29]

In response to the *SWANCC* opinion, the George W. Bush administration proposed a modified rule that would have removed federal protection from about a fifth of the wetlands in the lower forty-eight states, as well as many headwater streams. The administration soon abandoned this idea after more than forty states, countless conservation organizations (including hunting and fishing groups whose members had largely supported Bush's election), and members of Congress weighed in against any regulatory rollbacks.[30]

In 2006, the uncertainty resulting from the *SWANCC* opinion was greatly magnified by a second case—this time involving the Corps' regulatory jurisdiction over wetlands adjacent to tributaries of navigable waters. *Rapanos v. United States* combined two similar cases brought by developers and was named after the more provocative character. John Rapanos was planning a shopping center for one of his properties in Michigan, when he was told by a state inspector that the site likely had wetlands that were "waters of the United States." Rapanos hired a consultant who concluded that about fifty acres fell within the regulatory jurisdiction. Outraged at the government *and* the consultant, Rapanos defiantly began filling in wetlands on his properties. A civil enforcement action by the government led to the Supreme Court case.[31]

Once again, the high court ruled against the Corps. Further adding to the confusion, the justices for the majority broke ranks by giving two very different opinions for their ruling against the Corps. The high court never resolved which one to follow. Justice Antonin Scalia and three other conservative judges contended that federal jurisdiction extends only to wetlands that have a "continuous surface connection" to a "relatively permanent" body of water that is connected to interstate navigable water. This extremely narrow interpretation of the "waters of the United States" went against long-standing regulatory practices. Moreover, the Scalia opinion neglects to recognize the importance of streams and rivers that only flow seasonally or after rain—about 60 percent of stream miles in the United States.[32]

In the second opinion, Justice Anthony Kennedy sided with the conservatives but disagreed with their reasoning. In Kennedy's view, a wetland that isn't adjacent to a navigable body must have a "significant nexus" to traditional navigable waters to be a "water of the United States." This was a much less strict interpretation than Scalia's, but "significant nexus" is hard to define.

The *Rapanos* case potentially affects other parts of the Clean Water Act that are tied to the definition of "navigable waters," including permits for point source discharges and application of total maximum daily loads to nonpoint source pollution. At issue is where does the federal government's authority to regulate water resources under the Clean Water Act give way to the authority of individual states and private property rights?

This jurisdictional storm was long in coming. The Clean Water Act has not been amended in any significant way since 1987. "It's been an absurd period not to have been fine-tuned and brought up to date," notes William Andreen, a well-known legal expert on the Clean Water Act.[33] Yet any effort to amend the act faces enormous headwinds. With no help coming from Congress, the Obama administration took up the challenge to define the "waters of the United States" using Justice Kennedy's less strict "significant nexus" criterion. This effort is referred to as the Clean Water Rule, or alternately, as WOTUS—Waters of the United States.

The science is clear—wetlands, ponds, headwater streams, and rivers form a continuum. But the question is, when does a hydrologic connection become "significant" for downstream water quality? The EPA took several years to build their scientific case to answer this question, including a 408-page report based on twelve hundred peer-reviewed studies.[34] In addition, the agency considered over one million public comments on a draft rule, as well as input provided through over four hundred meetings with diverse stakeholders.

On June 29, 2015, the Corps and EPA finalized the Clean Water Rule. The proposed rule included a number of measurable criteria to determine which streams, wetlands, and ponds are "waters of the United States," and therefore regulated. Tributaries to traditional navigable waters would be automatically covered if they have "features" of flowing water—in other words, a bed, a bank, and a high-water mark. This designation included everything from small ephemeral creeks to rivers discharging to the sea. This was the standard generally applied before the Clean Water Rule.

The bigger challenge was how to determine which wetlands and ponds that lack a direct surface connection have a "significant nexus" to the river network. To address this issue, the Clean Water Rule set distance limits. For example, waters located in the one-hundred-year floodplain and within fifteen hundred feet of the ordinary high water mark of a traditional navigable water would be automatically jurisdictional, while those in the one-hundred-year floodplain outside that limit would require a case-specific evaluation. Recognizing that some "isolated waters" can be environmentally important, the rule also identified specific types of waterbodies that require a case-by-case analysis. Among these are California vernal pools and prairie potholes.[35]

A feature with a name like "prairie potholes" sounds insignificant, yet these depressions (remnants of continental glaciation) comprise thousands of wetlands in the northern Great Plains. They provide vital habitat along the Central Flyway, one of the three main migration routes of North American birds. Sometimes called the "duck factory" of the midwest, the prairie pothole region supports more than half of our nation's migratory waterfowl. Despite their ecological importance, more than half of all prairie pothole wetlands have been drained and converted to agriculture.[36]

The Clean Water Rule protects the types of water historically covered by the Clean Water Act, while expanding jurisdiction in some areas and narrowing it in others. Overall, the EPA estimated that the rule would result in a slight reduction in waters protected compared to practices that were in place prior to *Rapanos*. It was designed to provide something that Republicans often say they want and that had been sorely lacking—regulatory certainty.

Surprisingly, Corps officials made recommendations for broader federal authority as the rule was being finalized. A technical analysis by the Corps estimated that, due to changes made unilaterally by the EPA, as much as 10 percent of wetlands that previously had been covered by the Clean Water Act would be excluded under the new rule. Some members of the Corps disagreed so strongly that they requested all references to the Corps be removed from the rule and its supporting documents.[37]

The Clean Water Rule placed no additional permitting regulations on agriculture, yet it was widely criticized by the agricultural community. The rule continued to exempt "normal farming and ranching" practices (such as plowing, seeding, and cultivating), but with 75 percent of all wetlands on private property, farmers and ranchers were concerned that the rule might constrain activities on their land. During his time as Oklahoma attorney general, Scott Pruitt led a multi-state lawsuit against the rule. With no restraint of hyperbole, he called it "the greatest blow to private property rights the modern era has seen."[38]

Farmers were particularly worried that agricultural drainage ditches and canals would be regulated under the rule. Reagan Waskom, director of the Colorado Water Institute, explained the concern:

> Western farms are laced with canals that provide critical irrigation water during the growing season. These canals and ditches divert water from streams and return the excess through a downstream return loop, which is fed by gravity. Because they are open and unlined, they also serve as water sources for wildlife, ecosystems and underground aquifers. And because they are connected to other water bodies, farmers fear they could be subject to federal regulation.[39]

The EPA stated that the rule does not regulate "most" ditches—but the word "most" was hardly reassuring. Only ditches that drain wetlands or replace natural tributaries would be regulated, but it was difficult to assure everyone that the EPA (or citizen lawsuits) wouldn't upend this pragmatic approach. The American Farm Bureau Federation developed a "Ditch the Rule" social media campaign, warning that "any low spot where rainwater collects, including common farm ditches" might be regulated. The campaign was wildly successful, prompting the EPA to start its own social media campaign, "Ditch the Myth," explaining that the regulation actually excluded more kinds of ditches than previous federal rules. However, the EPA was far outgunned in the messaging battle, because the rule was just too complicated to summarize on a bumper sticker.[40]

The intensity of the pushback was overwhelming. Within days after publication of the rule, more than one hundred plaintiffs representing thirty-two states, agriculture, and environmental groups (the latter argued the rule was too weak) filed legal challenges in eighteen federal district courts and eight federal appeals courts.[41] Veteran Justice Department attorney Steve Samuels had never in his thirty years experienced anything like it, calling it a "nationwide game of whack-a-mole."[42]

On August 27, 2015, the day before the rule was to take effect, the U.S. District Court in North Dakota (home to many prairie potholes and large farming and oil and gas interests) blocked it in thirteen central and western states. Six weeks later, a federal appeals court blocked the rule's implementation nationwide, pending further action by the courts. This legal limbo included uncertainty about which court (district or appeals) should hear the challenges. This is where matters stood when Trump took office.

A case in northern California became a further rallying point for the agricultural community. In 2012, John Duarte had plowed forty-four acres of land to plant wheat, tearing through seasonal wetlands. Although "established and ongoing farming activities" are exempt from regulation, the recently purchased land had not been farmed for nearly a quarter century. Duarte ignored warnings that he would face penalties without first obtaining a permit from the Corps. In 2016, a federal court found Duarte liable for damaging the wetlands. The government sought a fine of $2.8 million and tens of millions of dollars in mitigation expenses. Duarte reluctantly settled, agreeing to pay $330,000 in fines and another $770,000 to be spent on restoring wetlands on other properties in the Sacramento Valley. The agricultural community was outraged. In their view, environmental regulators were totally out of control, having forced a farmer to pay more than a million dollars for simply *plowing his field*. "We're not going to produce much food under those kinds of regulations," argued one of Duarte's lawyers.[43]

On February 28, 2017, just hours before addressing his first joint session of Congress, where he promised "to promote clean air and clear water," Trump signed an executive order to redo the Clean Water Rule. Employing his own hyperbole, Trump lashed out that, "The EPA's regulators were putting people out of jobs by the hundreds of thousands . . . treating our wonderful small farmers and small businesses as if they were a major industrial polluter. They treated them horribly. Horribly." He even claimed that people have to worry about getting hit with a huge fine if they fill in a puddle.[44]

Trump's executive order had no legal significance—it was pure grandstanding. He could have accomplished the same thing by tweeting or telling Pruitt to start the legal proceedings. Several months later, Pruitt announced a proposal to repeal the WOTUS rule and plans to replace it with one that interprets "navigable waters" in a manner consistent with Judge Scalia's opinion—relatively permanent bodies of water and wetlands directly connected to navigable rivers. This replacement must go through the same public process that was used to develop the original rule: public comments, hearings, and agency response to the comments. Any replacement has to be justified by the law, legal precedent, and available evidence. There is also the ticklish problem that the Clean Water Rule had been built on considerable evidence from an extensive rule-making process that documented the critical importance of small streams and wetlands to the health of large rivers, lakes, and estuaries.[45]

In February 2018, the Trump administration officially suspended the Clean Water Rule for two years, while they worked to replace it. A cascade of lawsuits ensued. Then the unexpected happened. In August 2018, a district court judge in South Carolina breathed new life into the Clean Water Rule by blocking the Trump administration's suspension. The judge ruled that the government had failed to comply with the Administrative Procedure Act, having provided no "reasoned analysis" for suspending the rule and no "meaningful opportunity" for public comment. What this meant was that the Clean Water Rule was now law in twenty-six states, while remaining suspended in the twenty-four states where other federal district court judges had officially stayed it. The agricultural community was quick to respond. Scott Yager, chief environmental lawyer for the National Cattlemen's Beef Association, lamented that WOTUS was "back from the dead in 26 states, creating a zombie version of the 2015 rule that threatens the rights of farmers and ranchers across the country."[46]

In December 2018, the Trump administration officially proposed a replacement. This proposal would only protect wetlands with continuous surface-water connections to other protected streams and waterways, thereby slashing the number of wetlands protected by around half. The proposed rule also would exclude ephemeral streams—those that flow only after a rain or during a snowmelt. Such streams constitute a major part of the country's river systems, particularly in the western half of the United States. Justice Kennedy's opinion had explicitly rejected this idea. Requiring a continuous flow of water, he wrote, "makes little practical sense in a statute concerned with downstream water quality." Under that approach, "The merest trickle, if continuous, would count as a 'water' subject to federal regulation, while torrents thundering at irregular intervals through otherwise dry channels would not."[47]

The Trump administration's effort to upend WOTUS has begun the next round of interminable lawsuits and controversies that somehow have to untangle complex science, legal issues, and property rights. Neither Obama's Clean Water Rule nor the Trump administration proposal are completely defensible. The Clean Water Rule relied too heavily on hard to defend distance-based criteria, whereas the Trump administration proposed replacement arbitrarily reverses decades of regulatory norms and scientific consensus about the connectivity of waters and ecosystems. With the EPA now approaching its fiftieth anniversary, the fundamental question of what waterbodies are jurisdictional under the Clean Water Act remains unresolved.

Part III

Air Pollution and Climate Change

6

A NEVER-ENDING BATTLE

Discontent is the first necessity of progress.
—Thomas Edison

Los Angeles' first big "smog attack" came in the middle of World War II, when thick fog suddenly invaded the city.[1] Visibility was reduced to just a few city blocks. Eyes were stinging and noses were running. At first, people thought the Japanese were attacking with chemical warfare. The next day, city officials closed Southern California Gas Company's Aliso Street Plant that spewed noxious gases over a five-mile radius, thinking this was the source of the problem. Pollution controls were installed on the plant, but Los Angeles' smog attacks continued to worsen.[2]

By the early 1950s, Angelinos' eyes were irritated on a regular basis, headaches had become a fact of life, bronchitis and asthma were on the rise, and people were starting to forget what the sky looked like. California officials banned burning coal and fuel oils for industrial purposes, but the brown cloud over the Los Angeles basin just kept getting worse. By this time, smog was invading cities across the country, but Los Angeles was the undisputed Smog Capital of the United States.

In the early 1940s, Los Angeles had more than a million cars on the road. A decade later, there were over two million. Amazingly, no one was blaming the automobile. The thought *had* crossed people's minds, but auto exhaust was colorless, and smog was brown. There didn't seem to be a connection. It made a lot more sense to blame the city's geography. The Los Angeles basin is surrounded on three sides by mountains. Bad air gets trapped.

The smog mystery was finally solved by Arie Haagen-Smit, a chemist at the California Institute of Technology. It turned out that auto emissions, when exposed to sunlight, produce secondary pollutants—the brown stuff you can see that's known as smog. It came as quite a surprise to discover that Los Angeles' severe smog problem wasn't just because of all those cars—it was also because of all that California sunshine.

Haagen-Smit had discovered the problem, but that didn't mean politicians and the auto industry were lining up to congratulate him. Stanford Research Institute (funded by the oil industry at the time) sent someone to Caltech to discredit Haagen-Smit's findings, as well as his reputation. The only thing this accomplished was to motivate Haagen-Smit to double-down on his research efforts. Within a few years, there was no debate among scientists—cars were the smog culprit. While the number of cars continued to explode nationwide, there were no regulations on the auto industry to control emissions, or interest in doing so.[3]

In 1963, Congress passed the first (and largely unknown) Clean Air Act, which provided funds for federal research and assistance to states for starting air pollution control programs. A few years later, the Air Quality Act of 1967 gave the federal government limited enforcement procedures involving interstate air pollution. The key words here are *limited* and *interstate*. These efforts barely made a dent on the foul air Americans were breathing.

In 1970, Congress finally got serious about air pollution control and passed the Clean Air Act amendments, known today simply as the Clean Air Act. Among its major provisions, the act required a dramatic 90 percent reduction in emissions of carbon monoxide and hydrocarbons from new automobiles by 1975.[4] The act gave automakers a temporary escape clause for meeting the tougher standards, if they could prove they had made "good faith" efforts to control emissions. In such a case, the U.S. Environmental Protection Agency (EPA) administrator could grant a one-year delay. This marked the beginning of the tug of war that continues to this day between the EPA and the auto industry.

In 1972, automakers claimed they had made a good faith effort and applied for the delay. Meanwhile, a few smaller companies had made some impressive progress with a device they called the catalytic converter. While the automakers claimed that catalysts were plagued by problems, the catalyst companies insisted that the problems had been solved. Questions also arose about the automakers' procrastination in testing improved catalytic models. [5]

The EPA conducted a three-week-long public hearing during which forty-four witnesses exhausted every possible side of the problem. As the hearing progressed, it became obvious that the auto industry was much closer to achieving the 1975 standards than the EPA had been led to believe. William Ruckelshaus was in the hot seat, having to make a major decision that could affect the economy and jobs versus public health and confidence in the fledgling EPA. For two weeks, it looked like the decision could go either way. A turning point came during the third week of the hearing when David Hawkins, a young attorney for the Natural Resources Defense Council, carefully explained the lack of firm evidence supporting the automakers' contention that the standards couldn't be met. Hawkins warned that if the EPA granted a suspension based on the record that had been established, it would destroy the agency's ability at any later time to insist that the industry push itself harder to meet pollution control requirements.

After much internal deliberation, Ruckelshaus denied the extensions. The news hit the auto industry like a bomb going off. There was other fallout. The significance of the EPA's strict handling of the auto industry was not lost on the rest of American business. As a result, the EPA's 1972 auto decision remains a momentous event in the history of the environmental movement. In the euphoria following this decision, it seemed like the smog problem had been solved.

Less than a year later, the U.S. Court of Appeals overturned the EPA decision. The reasons given were that economic factors had to be given greater weight, and the evidence didn't justify denying the applications. The court directed the EPA to hold another public hearing. While impressive progress had been made with the catalytic converter in the interim, the automakers argued that they needed the delay because of the long lead times required to iron out all the wrinkles and gear up for mass production. This time, the EPA granted the one-year suspension.

Another event complicated the EPA's position. In October 1973, the Arab members of the Organization of Petroleum Exporting Countries placed an oil embargo on the United States as retaliation for supplying Israel with arms during the Yom Kippur War two weeks earlier. As oil suddenly came to a trickle from the Middle East, the price quadrupled. The oil embargo affected almost every American. Gasoline was rationed. Huge lines formed at the pumps. Many stations ran out of gas. The embargo continued for a year and a half.

When the embargo hit, a full year's production of 1973 cars was on the road. They had been subject to more stringent emission controls than earlier models, which had hurt their gasoline mileage. As fuel shortages drove up the price of gasoline and reduced mobility, car buyers made fuel economy one of the main factors in their choices. As the national economy began to slide into a recession, concern over the impact of federal requirements on the auto industry intensified. New car sales dramatically fell, resulting in layoffs of autoworkers. The combined effect of these developments shifted the public's attention from reducing auto emissions to fuel economy, cost, and jobs. In 1974, Congress granted an additional year's delay before the EPA's strict emissions standards would take effect. Other extensions were granted later. Because of these postponements, most cars manufactured through 1977 were emitting about four times as much pollution as the standard set by the 1970 Clean Air Act. The 90 percent reduction was finally achieved in 1981.[6]

A major side benefit of catalytic converters resulted because leaded gasoline "poisons" catalysts and they stop working. The EPA began its lead phase-out effort by proposing limits on the amount of lead that could be used in gasoline. The Ethyl Corporation, a joint enterprise of GM, DuPont, and Standard Oil of New Jersey, denied any harm to public health and fought restrictions on their highly profitable product. Removing lead from gasoline wasn't fully accomplished until 1996, but it ranks as one of the EPA's most significant accomplishments. [7]

Following these first attempts, auto emission and fuel economy standards have undergone additional tightening, yet many parts of the country still fail to meet air quality standards. Since 2010, standards for greenhouse gases have been added to auto emission standards, but it's a never-ending effort at playing catch-up. In recent years, the transportation sector (cars, trucks, trains, ships, and airplanes) has outstripped fossil fuel plants as the largest emitter of carbon dioxide in the United States.

During Obama's first term, the administration brokered an agreement with the auto industry to increase vehicle fuel efficiency standards and set the first ever carbon limits on cars and light trucks. By 2025, the goal was to nearly double the average fuel economy of new cars and light trucks to about fifty miles per gallon (generally fewer miles per gallon in real-world driving), and cut greenhouse gas emissions by half. These ambitious goals were the single biggest step the United States had taken to combat climate change. Other benefits included a cut in oil consumption, cleaner air, and saving money at the pump. The fuel efficiency standards would be achieved through a combination of electric cars, hybrids, and advances such as use of lighter materials and more efficient engines and air conditioners. It was estimated that this would add eighteen hundred dollars to the cost of a vehicle in 2025 but be more than offset by a vehicle-lifetime fuel savings of fifty-seven hundred dollars or more. [8] Better fuel economy also would mean more jobs. As Americans spend less money on gasoline, they will spend more in other parts of the economy, generating new jobs in sales, services, and manufacturing. The Union of Concerned Scientists estimated that by 2030 the standards would result in 650,000 new jobs. [9] Automakers, the United Auto Workers, state regulators, and environmental organizations were all onboard.

The problem was, this agreement with the automakers was brokered in 2009, when gas was expensive, and people were buying smaller, energy-efficient cars. Hybrids were selling like hotcakes. Obama also had considerable goodwill with the auto industry after having taken the bold (and highly controversial) steps to bailout General Motors and Chrysler during the Great Recession that hit in 2008. But starting around 2013, gas was cheap again and Americans once more wanted big sport utility vehicles, pickup trucks, and minivans (categorized as "light trucks" by the EPA). These vehicles consume more gasoline per mile, and many of them pollute three to five times more than small, energy-efficient cars.[10] Today, "light trucks" account for more than half of new passenger car sales. Appreciation of how Obama had gone out on a limb to save the auto industry also faded quickly.

When automakers signed on, they insisted on a mid-term review of the standards for model years 2022 to 2025. This review would consider a wide range of factors, including technology development and deployment, fuel prices, safety impacts, electric vehicle use, and employment impacts. A decision would then be made whether to retain, tighten, or loosen the standards. The Obama administration completed this review one week before Trump took office, concluding that the 2022–2025 standards can be met with only a slight modification on the mileage goals. Margo Oge, who directed the EPA's Office of Transportation and Air Quality for almost two decades, said the decision "was made on sound science and thousands of man hours of analysis."[11] The Alliance of Automobile Manufacturers (representing a dozen major car manufacturers) had a completely different view and encouraged Pruitt to rescind it. The industry claimed that over a million jobs were at stake.[12] This hyperbole is reminiscent of the public hearing with Ruckelshaus over forty-five years earlier, when automakers claimed that proposed emissions standards would destroy the industry. This time, however, the automakers were willing to negotiate. Mitch Bainwol, chief executive of the Alliance of Automobile Manufacturers, insisted that the car companies simply want a rational, predictable, stable policy. "We will get to the Obama numbers. We will get beyond the Obama numbers. The question is when and how," he said.[13]

With Trump's election, automakers had a ready champion for rolling back fuel efficiency and emissions standards, yet the problem was that such actions might not be in the automakers' best interests. Lowering standards will undermine the competitive edge U.S. manufacturers need in the global auto industry. China, India, and many European countries have announced even more stringent emission reduction goals and have embarked on an ambitious transition to electric cars.

There's another problem with rolling back standards. Under long-standing waivers that date back to the 1970 Clean Air Act, California is not required to go along with the EPA's auto standards—the state can, and does, set stricter standards. Making things even more complicated, thirteen other states and the District of Columbia have adopted California's standards. And so, the auto industry is caught in a bind. They want Trump to loosen national standards, but they would still have to meet the stricter standards for about 40 percent of the auto market. Instead of one size fits all, the auto industry would be getting into the business of custom designing their vehicles.

In April 2018, the Trump administration announced that it would soon be taking steps to roll back the Obama fuel efficiency standards. The administration also threatened to take away California's waiver. These proposals were well beyond what the auto industry wanted. Their ultimate nightmare was a protracted legal battle with an uncertain outcome. Rather than change the 2025 thresholds, the automakers sought more options for meeting them. "We support increasing clean-car standards through 2025 and are not asking for a rollback," wrote Ford's chairman and chief executive officer.[14] The Alliance of Automobile Manufacturers urged the White House to cooperate with California officials, saying "climate change is real."[15]

Pruitt ignored these entreaties. He planned to announce the proposed rollback with much fanfare at a dealership in suburban Virginia, but no dealership wanted to be associated with the announcement. Pruitt ended up making brief remarks at EPA headquarters, with limited media access.[16]

In August 2018, the Trump administration proposed the "Safer Affordable Fuel-Efficient (SAFE) Vehicles Rule," which would freeze fuel efficiency standards for automobiles at 2020 levels through 2026. The administration claimed this proposal would prevent more than twelve thousand road fatalities in comparison to the Obama plan.[17] One argument went like this: As people spent less on gas with more fuel-efficient cars under the Obama plan, they would spend more time on the road, therefore increasing fatalities. The Trump administration also claimed that higher prices for cars under the Obama plan would keep motorists in older, less safe cars. Behind the scenes, the EPA and Department of Transportation, which have joint jurisdiction over the clean car rules, clashed over the estimates and approaches. In a memo, the EPA called the Department of Transportation model "indefensible" and based on "unrealistic" assumptions. Even Acting EPA Administrator Andrew Wheeler questioned the auto fatality numbers. His concern was that, if they were proven faulty, it would undermine the legal case for the rollback. But he was overruled by the administration.[18]

The Trump administration also challenged the right of California and other states to set their own tailpipe standards. California Air Resources Board Chairwoman Mary Nichols had signaled a willingness to discuss altering the state's auto rules in the near term, if the administration agreed to support the efficiency targets further into the future.[19] "The backup plan is divorce," she announced after meeting with Wheeler. "I don't mean we're going to secede from the Union. We will reassert our Clean Air Act authority and move forward with our program, possibly with some improvements."[20] To the consternation of the auto industry, the Trump administration later cut off talks with California over the fuel efficiency standards and proposed to withdraw its waiver.[21]

In an interesting development, General Motors responded to the proposed SAFE Vehicles Rule by pushing for a National Zero Emissions Program. The proposal, modeled after California's Zero Emissions Vehicle Program, could add more than seven million electric vehicles to U.S. roads by 2030 and reduce carbon dioxide emissions by 375 million tons between 2021 and 2030. In addition to environmental considerations, GM's national program would aim to "preserve U.S. industrial leadership for years to come."[22]

In December 2018, a group of eleven scientists wrote a hard-hitting critique in *Science*, concluding that the Trump analysis "has fundamental flaws and inconsistencies, is at odds with basic economic theory and empirical studies, is misleading, and does not improve estimates of costs and benefits of fuel-economy standards beyond those in the 2016 [Obama] analysis."[23]

This same month, the *New York Times* revealed that refiners and other oil industry groups had been running a stealth campaign to roll back the standards. Cars use a quarter of the world's oil, and less-thirsty vehicles mean lower gasoline sales. In a remarkable statement, the oil industry argued that the United States is so awash in oil that it no longer needs to worry about energy conservation. Furthermore, "unelected bureaucrats" shouldn't dictate the cars that Americans drive. Oil industry groups even drafted legislation for states supporting this position and took out ads with messages like: "Support Our President's Car Freedom Agenda!"[24] Acting EPA Administrator Wheeler supported the oil industry view, saying that any government effort to encourage people to use electric vehicles was "social engineering."[25]

The Obama administration also set greenhouse gas and fuel efficiency standards for heavy-duty trucks and other large vehicles. While these represent only 4 percent of all vehicles on the roads, they generate 20 percent of the carbon pollution produced by the entire transportation sector.[26] Truck manufacturers fought tighter fuel economy standards, but truckers, who benefit from higher efficiency standards, were supportive. Shippers like FedEx also were on board.

In the final months of the Obama administration, the EPA closed a gaping loophole that allowed trucking companies to avoid regulations that applied to a new truck by installing an old engine into a new truck body (a so-called glider). These much cheaper glider trucks (also known as "zombies") spew up to forty times more nitrogen oxide and at least fifty times more fine particulates than new trucks. While virtually all truck and engine manufacturers were in favor of closing this loophole (by limiting the number of gliders per manufacturer to three hundred per year), Pruitt sought to bring the loophole back. Fitzgerald Glider Kits, the company that would benefit the most from the loophole, had met privately with Pruitt and hosted a campaign event for Trump at one of their dealerships. A study supporting the gliders case by Tennessee Tech University (paid for by Fitzgerald) was disavowed by the university as seriously flawed.[27] Not to be deterred, Pruitt issued a memo on his last day in office that the EPA would *not* limit sales of these super-polluting trucks. The DC Circuit Court intervened the following day, the EPA soon backed off, and the regulation limiting gliders remained in place.[28]

In November 2018, in a surprising move by the Trump administration, the EPA announced an initiative to decrease nitrogen oxide emissions from heavy-duty trucks. The initiative was a carry-over from the Obama administration. Acting Administrator Andrew Wheeler told reporters: "We are under no regulatory or court order requirements to launch this initiative. We are doing it because it's good for the environment"—as though this was something remarkable for the EPA to do.[29]

For almost a half-century, the EPA has been dueling with the auto industry to improve air quality. The EPA and the industry have had starkly different views of what is technically and economically feasible. Nonetheless, pressure from the EPA has caused the industry to get the lead out—in more ways than one. While the global auto industry is committed to fuel efficiency, reducing emissions, and transitioning to electric cars, the U.S. auto industry continues to churn out fleets of gas guzzlers to meet consumer demands. In positive developments, the tables appear to be turning somewhat with the U.S. auto industry pushing the EPA to more aggressively promote electric cars, and several automakers recently announced a compromise agreement with California on fuel economy standards. The oil industry is simultaneously pushing back against any move toward electric vehicles or better fuel economy. Reining in both industries remains a never-ending battle.

7

COSTS, BENEFITS, AND POLITICS

It is difficult to get a man to understand something when his
salary depends on his not understanding it.
—Upton Sinclair

Nothing that the U.S. Environmental Protection Agency (EPA)
does has had a greater impact on human health than air pollution
control. And as we've just seen with autos, it hasn't come easy. The
battle lines are firmly entrenched between the real (or perceived)
effects on jobs and the economy versus the more intangible benefits
of environmental and public health protections. From almost day
one, the costs and jobs faction has argued, "If it's so bad, where are
the bodies?" The benefits of environmental protections are extremely difficult to visualize on a day-to-day basis—and have nothing to
do with the equivalent of a bomb going off. The Clean Air Act is
one of the most complex statues ever passed by the U.S. Congress.
It attempts to take into account the myriad complications and longer-term risks from polluted air—to people and to the environment.

One of the most serious air pollution crises in the United States is
now largely forgotten, though it occurred just decades ago: acid
rain. Air pollution in highly industrial areas has long been known to
cause acid rain. Upon observing the effects of acid rain on vegetation around a copper mine in Sweden, the eighteenth-century biolo-

gist Carl Linnaeus lamented: "Never has a theologian described a Hell so dreadful as what is seen here!"[1]

In the 1960s and 1970s, acid rain reached the tipping point over a broad swath of the northeastern United States. In 1963, scientists discovered that rainwater in the Hubbard Brook watershed in the White Mountains of New Hampshire had a pH of four or less— compared to typical pH values of five to 5.5. The difference may seem slight, but it translates into ten to thirty times more acidity than normal. One sample measured 2.85—about the same acidity as lemon juice.[2]

By the 1970s, the pervasiveness and effects of acid rain were well recognized by scientists in the United States, Canada, and northern Europe. Two of the Hubbard Brook scientists—Gene Likens, an ecologist at Cornell University, and Herbert Bormann, a plant ecologist at Yale—brought the problem into the public's awareness in a 1974 article in *Science*. Based on a decade of study, Likens and Bormann declared unequivocally that "acid rain or snow is falling on most of the northeastern United States."[3]

Acid rain has multiple harmful effects on natural systems. It releases aluminum from soils to waterbodies, killing fish and other aquatic life. By flushing essential nutrients from soils, acid rain stunts forest growth. Stressed trees are vulnerable to insect attack and disease. The harmful effects of acid rain are amplified in mountainous areas with shallow soils. It also damages monuments and buildings.

The main source of acid rain was traced to sulfur dioxide and nitrogen oxide emissions from factories and fossil fuel power plants. These chemicals react in the atmosphere to form mild solutions of sulfuric and nitric acid. And it's far from a local problem. Much of the acid rain in the northeastern states was from emissions in the midwest and Great Lakes region, as winds blew over long distances. Acid rain from U.S. sources was also damaging Canadian forests.

Ironically, the problem was exacerbated by industry's attempts to meet air pollution standards. Their solution was to build ever higher smokestacks to disperse pollutants away from local areas, boosting long-range transport. In 1969, less than a dozen smokestacks were over five hundred feet tall; a decade later, there were 180.[4] Acid rain was also an unintended consequence from installing scrubbers in smokestacks that removed particles that previously had neutralized some of the emissions' acidity. And so, tall smokestacks and particle removal helped transform a local soot problem into a regional and international acid rain problem.

The discovery of acid rain in the United States was followed by almost two decades of denial and debate. To raise public awareness, the National Audubon Society launched a citizen science initiative in 1987 patterned after its century-long Christmas Bird Count. Some 280 volunteers across the country collected precipitation samples, measured the pH, and reported these to a central database, as well as garnered local media attention.[5] Finally, the 1990 amendments to the Clean Air Act gave the EPA authority to regulate sulfur dioxide and nitrogen oxides from fossil fuel power plants. These amendments included several progressive and creative approaches for achieving air quality goals, most notably the cap and trade pollution reduction program for sulfur dioxide emissions.

Cap and trade, one of those obvious solutions in hindsight, is a practical and effective approach for dealing with serious air pollution problems. The EPA first sets a pollution limit for an area's total emission budget. Each polluter/plant is granted the right to emit a certain amount of the total allowable emissions. If a plant spews off less than their permissible amount, it can sell the difference to a dirtier plant. This allows the buyer to stay in business, and the seller to profit from operating a cleaner plant. Over time, the total amount of pollution that is allowed diminishes. Owners of dirtier plants are obliged to steadily improve their technology, and owners of cleaner plants have economic incentives to keep improving. In theory, polluters who can reduce emissions most cheaply will do so, achieving the emission reduction at the lowest cost to society. And when all else fails, plants that release more pollutants than allowed face stiff monetary penalties. Cap and trade provides the private sector with flexibility in how they are going to reduce emissions, while also stimulating technological innovation and economic growth.

Cap and trade represents a novel approach to cleaning up the environment by working with human nature instead of against it. People resent being told what to do but like to turn a profit by being smarter than the next person. Initially, the idea was not well received by many in industry, as well as by many environmentalists, some of whom called it a "license to kill." A push by President George H. W. Bush and EPA Administrator William Reilly brought it to fruition.

The EPA's approach for tackling acid rain is one of the agency's most remarkable success stories. Despite the doomsday warnings from power industry officials that these regulations would cause a spike in electricity prices and lead to blackouts, acid rain levels have been substantially reduced, and electricity prices have increased along the usual trajectory. Dealing with acid rain was also the EPA's first high-profile international challenge. It remains one of the agency's primary legacies and is a potential approach to combat a much larger international challenge—greenhouse gas emissions.

Acid rain doesn't conform to state boundaries, illustrating why you can't turn air pollution control completely over to the states. Upwind states can reap the benefits of factories with belching smokestacks, whereas downwind states, through no fault of their own, are unable to meet EPA standards. The late New Jersey Senator Frank Lautenberg repeatedly pointed out, "On some days even if we shut down the entire state, we would be in violation of some health standards because of pollution coming over from other states."[6] Air pollution in the form of regional haze also affects visibility in many national parks, including the Grand Canyon, Yosemite, the Great Smokies, and Shenandoah National Parks.

The Clean Air Act has a "good neighbor" provision that requires states to limit emissions that cause air quality problems in downwind states. However, regulating interstate air pollution has been highly controversial for decades. There is, of course, the issue of the federal government overstepping the authority of states. But the debate also centers around the difficulties of nailing down the sources of air pollution in any given location. Unlike water in rivers, air pollution doesn't move along simple flow paths. This makes modeling downwind emissions over large areas an imprecise science, opening the door for bad-neighbor states to challenge requirements for emission reductions. The latest approach, the Cross-State Air Pollution Rule (CSAPR; pronounced "Casper") was finalized in 2011, struck down by the U.S. Court of Appeals in 2012 (with Brett Kavanaugh writing the majority opinion), and then reinstated by the U.S. Supreme Court in 2014. "EPA must have leeway in fulfilling its statutory mandate," the high court ruled, as long as the agency reasonably balances the possibilities of undercontrol and overcontrol.[7] CSAPR requires certain states in the eastern half of the United States to reduce power plant emissions that cross state lines and contribute to smog and particulate pollution in downwind states. Emission trading programs for sulfur dioxide and nitrogen oxide build on the lessons from acid rain.

PRINCIPAL POLLUTANTS

The Clean Air Act authorizes the EPA to set national air quality standards for six principal pollutants: carbon monoxide, particulates (tiny particles in the air), ground-level ozone, sulfur dioxide, nitrogen oxides, and lead. To achieve its goals, the EPA addresses both "mobile" sources (cars, trucks, and airplanes) and "stationary" sources (power plants and factories). Each state must develop an EPA-approved plan to describe how it will control air pollution. (Indian tribes can develop their own plans.) Many of these efforts focus on "nonattainment areas" that fail to meet the standards set for the six principal pollutants. Of these pollutants, ground-level ozone and fine particulates present the most serious and widespread health threats. To illustrate the face-off among the costs, benefits, and politics of their regulation, let's take a closer look at the ozone standard.

Ground-level ozone, the primary component of smog, can trigger coughing and wheezing even in healthy adults. Paul Billings, with the American Lung Association, compares breathing ozone to a "sunburn of the lungs."[8] Children, the elderly, and those already suffering from respiratory illnesses like asthma are especially vulnerable to ozone pollution. Nearly thirty million people in the United States have been diagnosed with asthma.[9]

Ozone is one of those curiosities that, depending on where it's located, is a very good thing or the reverse. "Good" ozone in the upper atmosphere (stratosphere) shields the Earth from most of the sun's ultraviolet light—making life possible, and on a day-to-day level, protecting us from skin cancer and cataracts. Meanwhile, ground-level ozone is a potent respiratory hazard and pollutant that causes human health problems and damages plant tissue in forests and crops. It's only relatively recently in history where human "advancements" have threatened the good ozone and the bad ozone has threatened us. You could say the Industrial Revolution kicked it off, but it wasn't until automobiles reached a certain critical mass for the problem to be recognized. Eventually, the EPA was given the task of slaying the ground-level ozone dragon. It's been a major fight ever since.

Ozone pollution arises from volatile organic compounds (VOCs) and nitrogen oxides, and there are literally hundreds of millions of sources. Sources of VOCs include motor vehicles, the chemical and petroleum industries, dry cleaning, and the widespread use of solvents. As cars have become cleaner, consumer products such as cosmetics, paints, and indoor cleaners are an increasingly dominant source of VOC emissions. Almost every consumer product that has a smell emits VOCs. Chris Cappa, a researcher at the University of California at Davis, explains, "Say somebody is inside using perfume, cologne. That smell eventually dissipates. And the question is, where did it go? Those odors dissipate because it's basically getting moved outside."[10]

In nonattainment areas, specific controls might be set on smaller pollution sources, such as gas stations and paint shops. Metropolitan areas with the worst ozone pollution are required to use reformulated gasoline, which burns more thoroughly and evaporates slower to reduce hydrocarbon emissions. Reformulated fuels have the added advantage of containing much lower concentrations of toxics, such as benzene. The EPA also encourages use of alternative fuels, such as natural gas, biodiesel, and electric cars. If the EPA determines that a state is flouting air quality goals, it has the power to withhold federal highway funds or impose other sanctions—although it rarely does so.

On a national level, ground-level ozone levels fell 32 percent from 1980 to 2017, yet meeting the standard continues to elude many areas, including much of California, the Northeast Corridor, and major cities.[11] And even if the elusive goal is attained, there's a scientific consensus that the current ozone and particulate standards need to be lowered still further to protect public health, especially for young children. Anticipating this kind of flux, the Clean Air Act requires the EPA to review the standards for the six principal pollutants at five-year intervals and revise them, as appropriate. The ozone standard illustrates these challenges.

In 1997, the Clinton administration significantly tightened the standards for ozone (and particulates). This decision came after a fierce behind-the-scenes battle between EPA Administrator Carol Browner and Clinton's economic advisors. Industry groups warned that tightening the standards would deal a "crushing blow" to the economy. In announcing his decision, Clinton declared, "I approved some very strong new regulations today that will be somewhat controversial, but I think kids ought to be healthy." According to Browner, the revised standards would "provide new health protections to 125 million Americans, including 35 million children." The EPA gave states and cities substantial flexibility in deciding how to reach the new goals.[12]

Over the next decade, evidence continued to mount that the ozone standard should be tightened still further. In 2008, the Bush administration proposed to change the standard from eighty-four (the 1997 level) to seventy-five parts per billion.[13] This was significantly weaker than the sixty to seventy parts per billion standard that had been unanimously recommended by the EPA's Science Advisory Board. Public health and environmental groups took the Bush administration to court, arguing that the revised standard was politically motivated and not protective enough of human health. In 2009, when Lisa P. Jackson was appointed EPA administrator by

Obama, she asked these groups to hold their lawsuit in abeyance while she reconsidered the ozone standard. In the spring of 2011, Jackson decided that a standard of sixty-five parts per billion was necessary to protect public health with an adequate safety margin.

The ozone decision pitted Jackson, a Princeton-trained chemical engineer, against the White House chief of staff, William M. Daley, son and brother (respectively) of Richard J. and Richard M. Daley, the powerful Chicago mayors who long dominated the city's politics. Jackson was fully aware that her proposed standard would cause political heartburn at the White House, so she met with Daley multiple times before submitting it. Daley listened politely, but at one point asked, "What are the health impacts of unemployment?" To appease Daley and other critics, Jackson agreed to settle for a weaker standard of seventy parts per billion—the upper limit recommended by the EPA's Science Advisory Board. She also agreed to significant flexibility in compliance. Jackson thought she had a deal.[14]

Each incremental tightening of the ozone standard comes at a significant cost to business and government, as it can lead to more nonattainment areas requiring expensive pollution control efforts. As a result, the ozone standard became a symbol of what opponents called a "regulatory jihad" by the Obama administration. Industry lobbyists and Republicans in Congress identified it as one of their top targets. Local and state officials argued that they lacked the resources to enforce the new rule. Ground-level ozone had become such a lightning rod for opposition that some Democratic lawmakers cautioned Obama that the regulation would damage their re-election prospects. The impact would be felt heavily in the midwest and Great Plains states that would be critical electoral battlegrounds in 2012.[15]

Nevertheless, Jackson pushed onward, meeting with industry groups to make her case. "Lisa is very smart, cordial, friendly," R. Bruce Josten, the chief lobbyist for the U.S. Chamber of Commerce, said. "She listened to us, but then talked about how important it was to do this, the lung thing, the asthma thing, the kids' health thing. She felt it was important to go ahead. . . . The funny thing was nobody wanted to come right out and say, 'Are you guys thinking this through? Your boss is up for re-election next year.'"[16]

On the first of September 2011, Obama summoned Jackson to the White House. In a terse meeting, Obama informed Jackson that tightening the ozone standard would impose too severe a burden on industry and local governments during a time of economic distress. He told Jackson that she would have an opportunity to revisit the standard two years later—*if* they were still in office. "We are just not going to do this now," he informed her. When the White House announced the decision the next day, environmental and public health advocates were livid. Jackson reportedly considered resigning.

The retreat on the ozone standard was the first environmental decision of the presidential campaign season that was now fully underway. The White House announcement came barely an hour after a weak jobs report was issued by the Labor Department. Still reeling from the 2008 economic meltdown, the country was in the midst of an intensifying political debate over the impact of federal regulations on job creation.[17]

Lisa Jackson resigned at the end of Obama's first term. Her successor, Gina McCarthy, continued the ozone battle during Obama's second term—this time with the president's support. In October 2015, the EPA finalized revisions to the ground-level ozone standard, tightening it from seventy-five to seventy parts per billion. The Edison Electric Institute, a powerful trade group for utilities, concluded that a more stringent rule was inevitable and lobbied for the lesser of two evils—a seventy parts per billion level, compared to sixty-five. Part of the reason for this unexpected support was that its member utilities had already cleaned up their dirtiest coal plants.[18]

The EPA estimated that reducing ozone concentrations to seventy parts per billion would prevent 230,000 asthma attacks in children and 320 to 660 premature deaths annually when fully implemented in 2025. These figures didn't count California, which has longer to meet the standard because of its more extensive ground-level ozone problem. Los Angeles and California's San Joaquin Valley continue to have some of the worst air quality in the nation.[19]

The Clean Air Act prohibits the EPA from considering costs in setting standards for the six principal pollutants. This position was confirmed in a unanimous 2001 U.S. Supreme Court opinion, written by no less than conservative Justice Antonin Scalia.[20] The act simply directs the EPA to set the primary standard at a level necessary to protect public health, "allowing an adequate margin of safety." In the real world, however, the battle between business and the EPA mainly comes down to cost. The EPA prepares cost and benefit estimates for information purposes, as well as to comply with Office of Management and Budget requirements for economically significant rules. The agency estimated that the ozone standard of seventy parts per billion would produce benefits of three billion to six billion dollars a year while costing about $1.4 billion annually (excluding California).[21]

Preparing cost and benefit analyses is just one of many steps. Revisiting the standards for ozone, particulates, or other principal pollutants is a long and complicated undertaking. EPA scientists must first compile and summarize the scientific literature published since the last revision. For the 2015 ozone rule, the EPA reviewed more than one thousand scientific studies, covering such wide-ranging topics as the physics and chemistry of ozone in the atmosphere, environmental concentrations, toxicology, epidemiology, interactions with co-occurring pollutants, and so on. The EPA then prepared a risk and exposure assessment to identify exposure pathways, at-risk populations, and health endpoints. Historically, a panel of leading experts on the health and environmental effects of the pollutant under consideration evaluates the agency's work throughout this process.

If a revised standard is determined, the states and EPA identify nonattainment areas. State and local governments then have up to three years to produce implementation plans, which outline the measures that will reduce emission levels to attain the standard. After the plans are approved (which can take years), the actual attainment of the standard is allowed to stretch over a three- to twenty-year period, depending on the severity of the area's pollution. For the 2015 revised ozone standard, the EPA was required to designate nonattainment areas by October 2017. In June 2017, the EPA announced a one-year delay on the grounds that the agency didn't have enough information to identify nonattainment areas. This was pure nonsense. The states had submitted this information the previous fall, when Obama was still in office. [22]

Shortly after the EPA announced the delay, fifteen states and the District of Columbia filed a lawsuit, blasting the decision. Recognizing that it had no legal leg to stand on, the EPA backed down the following day. Pruitt tried to put a positive spin on the reversal: "We believe in dialogue with, and being responsive to, our state partners."[23] This was a rare early win for environmentalists under the Trump administration.

In November 2017, several weeks after the October due date, the EPA certified that 2,650 areas were in compliance with the ozone standard—but failed to identify any nonattainment areas, or release a timeline for doing so.[24] The agency was sued for the delay, and months later identified fifty-two areas in twenty-two states that did not meet federal ozone requirements. More than 124 million people live in these nonattainment areas and are breathing ozone above health protective levels.[25]

The next five-year reviews of ozone and particulates are currently underway. Once again, these are controversial as the Trump administration attempts to fast-track the process and reduce its scientific rigor.[26] The reviews rely heavily on recommendations by the Clean Air Scientific Advisory Committee (CASAC), whose current seven members are all Trump administration appointees. The chair is a consultant long funded by the fossil fuel industry and a skeptic of the links between air pollution and health. He questions approaches that use multiple lines of evidence in favor of considering only studies that pass a very narrow (perhaps unachievable) level of causality. In addition, the CASAC has always relied on a panel of experts for each pollutant under review, but these were disbanded by the Trump administration for expediency. Despite calls by numerous experts to reinstate them, EPA Administrator Andrew Wheeler claims that the review panels are not needed. In a rebuke of Wheeler, the current CASAC members have said they don't have sufficient expertise and need the panels.

MERCURY AND AIR TOXICS

Air pollution isn't just restricted to the six principal pollutants. Many other air pollutants are known, or suspected, of causing cancer or other serious health problems. One of the most troubling is mercury. Prior to regulation by the Obama administration, coal-fired power plants accounted for about half of U.S. mercury emissions.[27]

Mercury has some amazing and alarming characteristics. In the amazing category, it's the only metallic element that is liquid at room temperature (hence its use in thermometers). And unlike other common heavy metals, mercury readily vaporizes, making it transportable to any location. In the alarming category is the fact that it doesn't degrade over time and is one of the few toxic elements that biomagnifies in aquatic food webs, posing a threat to fish, wildlife, and humans as it works its way up the food chain.

Mercury released into the air from coal-fired power plants ends up in lakes and other waterbodies, many of which are located in relatively pristine areas. Once in water, mercury is commonly transformed into methylmercury—the form that readily moves up the food chain. Methylmercury is highly toxic to the nervous and cardiovascular systems. It is particularly dangerous to developing fetuses and infants. Even low exposure levels can impair cognitive development and IQ in this vulnerable population. What makes this risk even more insidious is that there's no evidence for a threshold of these effects—which means that any exposure to methylmercury is potentially a health risk. Yet, despite the hazards, controlling mercury emissions continues to be one of the most controversial air pollution debates.

Humans are exposed to mercury primarily by eating fish, and this is a widespread problem. About 80 percent of all fish consumption advisories in the United States focus on mercury contamination. These advisories have been issued in all fifty states—covering 16.8 million acres of lakes, 1.25 million river miles, and the coastal waters of twenty states.[28] Mercury concentrates in the muscle tissue of fish. Unlike PCBs, dioxins, and other organic contaminants that concentrate in the skin and fat, mercury cannot be filleted or cooked out of fish.

During its first two decades, the EPA's progress on toxic air pollutants (mercury, arsenic, cyanides, and so forth) was extremely slow, with regulations promulgated for only seven toxic pollutants. Unhappy with this slow pace, Congress took it upon themselves to compile a list of 187 hazardous air pollutants (air toxics) as part of the 1990 Clean Air Act amendments. The EPA was given ten years to set "maximum achievable control technology" standards for any source that emitted large amounts of these air toxics. Polluters then had three years, with a possible one-year extension, to meet the standards. The regulatory pace picked up considerably.

While industries from oil refineries to steel mills complied with the new rules, lobbyists for the electric utilities managed to get a delay built into the act. This was done by using the tried and true delay tactic of needing to study the problem. The EPA was required to first complete a study for Congress on whether regulation of air toxics emissions from power plants was really needed. To no one's surprise, the EPA determined that the regulation was "appropriate and necessary." This announcement was made in December 2000, at the end of the Clinton administration. A month later, George W. Bush came into office with support from the fossil fuel, electric power, and other high-polluting industries.

Bush's primary air pollution initiative, a so-called Clear Skies Act, sought to replace the EPA standards for mercury with a cap and trade system similar to that used to address the acid rain problem. The cap and trade system would allow power plants to avoid mercury emission controls if they obtained allowances from others who achieved lower pollution levels than required, or reduced emissions sooner than required. The success of cap and trade for acid rain made this proposal look good on paper, but there were a couple of shortcomings. A primary problem was that "hot spots" of mercury concentrations would continue in some waterbodies. The proposed cap also would lead to slower progress in mercury reduction than required by the 1990 amendments.[29] The state of New Jersey, along with others, challenged the cap and trade rules. In 2008, the DC Circuit Court of Appeals ruled in New Jersey's favor, and the incoming Obama administration began a more rigorous approach to lowering mercury emissions.

In 2011, the EPA finalized standards on mercury, arsenic, and other air toxics emitted by coal- and oil-fired power plants. Widely known as the Mercury and Air Toxics Standards (mercury rule for short, but also known as MATS), the goal is to prevent about 90 percent of the mercury in coal from entering the air. During the public comment period leading up to the standards, the EPA received almost a million comments. Industry claimed severe economic consequences, whereas environmentalists applauded the rule. When the standards were released, the president and chief executive officer of the National Association of Manufacturers warned that utility companies "will be forced to shut down power generation plants throughout the country, and the reliability of the power grid will be threatened." If that wasn't bad enough, he warned that "a jump in energy prices will have a devastating impact on companies of all sizes, harming their ability to create jobs, invest and grow."[30]

Despite these dire warnings, the Congressional Research Service found that the rule would have very little effect on electricity rates and reliability. The EPA projected an increase of 3 percent in the cost of electricity in 2015 due to the initial financial outlay in meeting the rule. This would fall to less than 1 percent by 2030. In addition, about 60 percent of existing coal-fired power plants were already in compliance, and less than *one half of one percent* of the nation's electric generating capacity (mostly coal plants over thirty years old) would likely be forced to shut down. Clearly, there was no threat of an energy Armageddon.[31]

The effect on jobs was muddier. Retirement of older plants and increased electricity costs for energy-intensive industries could lead to job losses. The EPA countered that jobs would be created for construction, installation, and operation of pollution control equipment. Not every utility objected to the mercury rule. Those that relied heavily on nuclear power or natural gas, or that had already invested hundreds of millions of dollars in upgrades, generally supported the rule.

Environmentalists saw the mercury rule as Obama's best leverage to reduce coal burning. Industry made it a poster child for the "war on coal." As Oklahoma attorney general, Pruitt joined more than twenty states in a lawsuit opposing the rule. In 2015, in a partial victory for opponents of the mercury rule, the Supreme Court left the rule intact, but sent the EPA back to the drawing board to reconsider the costs to industry.

The mercury rule is among the most expensive that the EPA has ever promulgated. Obama's EPA, however, found the costs to be justified in light of enormous health benefits. The agency estimated that for every dollar spent to reduce pollution, Americans would receive three to nine dollars in public health benefits. These arise in large measure because the equipment used to comply with the mercury rule also would reduce fine particulates in the air, leading to the annual prevention of eleven thousand premature deaths and 130,000 asthma attacks, among other health benefits. [32]

A positive indication of progress came in 2016 when scientists reported rapidly declining levels of mercury in Atlantic Bluefin tuna. [33] Long-standing public concerns about the connection between tuna and mercury date back to 1970, when a chemistry professor in New York City found high levels of mercury in a can of tuna that spurred a nationwide recall.

In 2018, Trump's EPA announced that it planned to reconsider the mercury rule. By this time, most power plant operators had complied by shutting plants down or retrofitting them. Desiring legal certainty, the electric power industry reversed its position and now urged the EPA to keep the rule. Like the auto emissions standards, it was another case of Trump's EPA proposing to roll back regulations well beyond what even industry wanted.[34]

On December 28, 2018, just hours before closing its doors for the year (as part of a partial government shutdown over Trump's border wall), the EPA announced that the mercury rule would remain in place as is. There was, however, a major catch. The EPA was no longer going to consider the public health benefits that came from reduction of pollutants other than those targeted. Neglecting these co-benefits sets a troubling precedent for protecting public health. Without including the co-benefits of fine particulates, the costs of the mercury rule would have exceeded the calculated benefits. Complicating matters, it's particularly difficult to put a specific dollar figure on some health benefits, such as those from avoiding lost IQ points in infants or other fetal harm from pregnant women eating mercury-contaminated fish. For that reason, the original mercury rule argued against using a strict cost-benefit analysis to decide whether the regulation should be imposed.[35] In addition to opposition to the EPA's move by health, business, environmental, and social justice groups, a bipartisan group of senators, including Senator Manchin from coal-rich West Virginia, wrote to Wheeler that they would "not support any efforts that might jeopardize" the mercury rule.[36]

SECRET SCIENCE ACCUSATIONS

In 2018, air pollution was on the front lines of yet another major attack on science. In April, Scott Pruitt held a press conference to announce a proposed rule-making called *Strengthening Transparency in Regulatory Science*. "The era of secret science at EPA is coming to an end," he announced.[37] In developing regulations, the EPA would no longer consider studies for which the underlying data were not publicly available. While this may sound like an honest approach, it was, in fact, a thinly veiled effort to undermine key studies that have helped justify stricter limits on air pollution, as well as some toxic chemicals. Health research often contains confidential personal information that is illegal to share. As such, the bill would prevent the EPA from using many of the best scientific studies.

In particular, the new policy was aimed at blocking the EPA from considering large epidemiological studies, such as those that have revealed the health dangers of fine particulates. Fine particulates readily find their way deep in people's lungs, where they can induce a wide spectrum of heart and lung problems that include a heightened risk of premature death. In the mid-1990s, two major epidemiological studies—known as the Harvard Six Cities and American Cancer Society studies—tracked the medical histories of thousands of people exposed to different levels of air pollution. These studies, based in part on confidential health information, found that exposure to even relatively low levels of fine particulates increased premature deaths.[38]

These two studies were pivotal in supporting the development of more stringent regulation of fine particles and suggest that the benefits of even tighter regulations would outweigh the costs. Burning coal and other fossil fuels is a major source of fine particulates. Consequently, studies linking health improvements with reduction in particulate emissions have been under attack by industry and Republican lawmakers for more than two decades. Adding to the controversies, Obama's proposed regulations to curb greenhouse gas and mercury emissions from fossil fuel power plants were supported, in large part, by the associated health benefits from reducing particulate emissions.

In an irony of ironies, Pruitt's unveiling of the EPA's proposed "transparency" rule was closed to the public and the press. The audience consisted of a who's who of climate deniers who had worked to weaken established climate science. In developing the proposed rule, the EPA also bypassed its own Science Advisory Board, which normally provides input on such a major planned action—in this case, an action that would allow the EPA to ignore peer-reviewed scientific studies. Nearly seventy medical societies and public health groups expressed opposition to the proposal.[39]

Standing next to Pruitt at the cloistered unveiling of the transparency rule was Representative Lamar Smith (R-TX), the preeminent climate-change denier in the House. As chairman of the House Science Committee, Smith had worked for years to promote bogus scandals against climate scientists and discredit evidence for human-caused climate change. The proposed regulation mirrored legislation long-championed by Smith. His bill, originally called the Secret Science Reform Act and later the HONEST (Honest and Open New EPA Science Treatment) Act, passed the House three times but was never taken up by the Senate. After the repeated failure of the bill to move forward, Smith's office had switched tactics by working with political appointees within the EPA to implement the idea through internal policy change. "It just keeps coming back in different forms. . . . It's like malaria. Or maybe herpes would be a better analogy," commented toxicologist Dan Costa, who had recently retired after leading the EPA's air research program for fourteen years.[40]

Most everyone agrees that increasing the public availability of scientific information for independent validation is a worthy cause. Academicians, scientific publishers, and funding agencies have worked in recent years to develop principles to encourage the ability for others to replicate studies. The long-term trend is for authors to supply access to data and analytical methods after their scientific findings are published. However, Pruitt's proposed approach of excluding peer-reviewed scientific studies because of confidential health records is beyond the pale. The EPA's rule-making must be based on the best information available. So-called reforms that are proposed by members of Congress and political appointees, with no input and against the advice of scientists and mainstream scientific organizations, are highly troubling. Fortunately, it appears unlikely that the proposed transparency rule will hold up in court. As Richard J. Lazarus, professor of environmental law at Harvard, noted after announcement of the proposed rule, the EPA would be "walking into a judicial minefield" by instructing its staff to ignore certain key studies during agency rule-making.[41] It also may violate the Toxic Substances Control Act, which requires the EPA to rely on "best available science" and "reasonably available information," taking into account the "weight of scientific evidence."[42]

It's not surprising that automakers, industries, and businesses resist air pollution regulations. Individual consumers also are affected. When the dashboard "check engine" light comes on, it often means a malfunction of a pollution control device. This can be unwelcome news for the car owner, because the fix may be expensive. Nonetheless, it is well-documented that clean air and a strong economy can go hand in hand. From 1970 to 2017, aggregate national emissions of the six common pollutants dropped an average of 73 percent. During this same period, the U.S. gross domestic product more than tripled.[43]

The Clean Air Act has averted an estimated 160,000 deaths and eighty-six thousand hospitalizations annually since 1990.[44] This may be the proverbial tip of the iceberg, as many more cases of respiratory and other health insults have been averted that do not require hospitalizations or even visits to the doctor's office.[45] A recent study found that thanks to the strides made in cutting air pollution, children's lungs in Southern California are 10 percent bigger and stronger today than they were in children twenty years ago.[46]

Among the EPA's most significant accomplishments was eliminating lead from gasoline. The health effects from exposure to lead range from behavior disorders and anemia to mental retardation and permanent nerve damage. Children are especially susceptible to lead's toxic effects on the nervous system, which can result in learning deficits and lowered IQ. In the late 1970s, 88 percent of American children had elevated levels of lead in their blood. By the mid-2000s, that number had dropped to less than 1 percent.[47]

The EPA also has been successful in addressing stratospheric ozone depletion. Unfortunately, the depletion that occurred before regulations took hold may take more than a half-century to completely heal—a warning signal about procrastinating on regulating greenhouse gases.[48]

Air pollution is the leading environmental cause of death worldwide. Considerable progress has been made, but many areas in the United States still fail to attain health-based goals, particularly in poorer communities. A recent study published by the National Academy of Sciences documented that non-Hispanic whites enjoy a so-called pollution advantage—they bear the burden of 17 percent less air pollution than is generated by their activities. Blacks and Hispanics, on the other hand, experience a "pollution burden," facing greater than 50 percent more exposure than is caused by their activities.[49] The need to address air pollution from numerous sources remains a never-ending imperative.

8

CLIMATE CHANGE

We, the people, still believe that our obligations as Americans are not just to ourselves, but to all posterity.
—Barack Obama, in his second inaugural address

It is unconscionable that the United States has simply walked away from its responsibility to people both at home and abroad, in the interest of short-term fossil fuel profits, and abandoned an agreement that was negotiated by more than 190 world leaders, over decades.
—Mary Robinson, former president of Ireland and UN special envoy on climate change [1]

In 1989, George H. W. Bush was the first president to pay a visit to the U.S. Environmental Protection Agency (EPA) headquarters. The occasion was the swearing-in ceremony for William K. Reilly as EPA administrator, the first "professional environmentalist" to head the agency. To counter the environmental hornet's nest stirred up by the Reagan White House, Bush had campaigned as "the environmental president." Speaking before some five hundred EPA employees, Bush emphasized, "I hope it is plain . . . to everyone in this room and around the country that among my first items on my personal agenda is the protection of America's environment." The EPA employees heartily applauded. [2]

Trump also visited the EPA headquarters shortly after his inauguration, but he came for an entirely different purpose—to deliver another slap in the face to the beleaguered agency. The email announcing Trump's visit arrived at lunchtime. "Our Big Day Today!" read the subject line of the message, which went to thousands of EPA employees. "This is an important moment for EPA," wrote Chief of Staff Ryan Jackson. He also cautioned that there was "limited space" to see Trump.[3] As energy industry representatives and their political friends filled the seats, most employees were shut out. EPA staff, those who could stomach the occasion anyway, watched over closed-circuit television as Trump declared the EPA's "war on coal" and its "attack on American industry" to be over.[4]

Trump sat at a small table where he signed an executive order instructing federal regulators to rescind key Obama-era rules curbing U.S. carbon emissions. "Come on, fellas. Basically, you know what this is," Trump gloated to the coal miners gathered around him. "You know what it says, right? You're going back to work."[5] This was the latest in Trump's never-ending series of photo ops.

Trump's strong embrace of coal was serendipitous. In May 2016, Trump addressed one of the largest rallies of his campaign in Charleston, West Virginia. With the front rows packed with mine workers, an official from the West Virginia Coal Association handed him a miner's hat. As Trump put it on, he gave the miners a double thumbs-up. "The place just went nuts, and he loved it," recalled Barry Bennett, a former adviser to Trump's presidential campaign. "And the miners started showing up at everything. They were a beaten lot, and they saw him as a savior. So, he started using the 'save coal' portions of the speech again and again."[6]

In 2014, Obama had pledged to reduce America's greenhouse gases 26 to 28 percent from 2005 levels by 2025. This goal was jointly announced in Beijing with China's President Xi Jinping, who set a goal for China's emissions to peak by 2030. The U.S. changes would be made through greater fuel efficiency for cars and light trucks, new rules governing emissions from coal-fired power plants, and limits on methane emissions from oil and gas wells. The Trump administration would fight all of these. There was no surprise here, given Trump's six-year string of more than one hundred tweets disparaging human-caused climate change, including his bogus claim that, "The concept of global warming was created by and for the Chinese in order to make U.S. manufacturing non-competitive."[7]

Within hours of Donald Trump's swearing-in as president, climate change (along with health care and LGBT rights) disappeared from the White House webpage. In its place was an *America First Energy Plan*, pledging to roll back "burdensome regulations," "embrace" oil and gas, and revive America's coal industry, "which has been hurting for too long." The White House boasted that lifting these and other restrictions "will greatly help American workers, increasing wages by more than $30 billion over the next 7 years."[8]

Asked later about why the Trump administration proposed slashing federal funding for climate change related programs, Mick Mulvaney, director of the Office of Management and Budget, responded: "We're not spending money on that anymore; we consider that to be a waste of your money to go out and do that."[9] Meanwhile, atmospheric carbon dioxide—the primary greenhouse gas that drives climate change—was at 407 parts per million (a level that last occurred over eight hundred thousand years ago).[10] This level was up from 384 parts per million a decade earlier and about 280 before the Industrial Revolution.

By the time of the Trump administration, virtually all Republicans were either silent on the topic or climate change deniers. "Most Republicans do not regard climate change as a hoax," claimed a campaign strategist for Senator Marco Rubio, but "it's become yet another of the long list of litmus test issues that determine whether or not you're a good Republican." This attitude is a recent development. In 2008, Senator John McCain ran for president as a candidate who had stood up to President George W. Bush and "sounded the alarm on global warming." In 2009, the House of Representatives narrowly passed a cap and trade bill (analogous to the one addressing acid rain) to regulate greenhouse gases, but the bill died in the Senate. It was Obama's first major legislative defeat. The tide was turning with groups like the Competitive Enterprise Institute and the

Koch Brothers' Americans for Prosperity waging an all-out campaign to elect lawmakers who were friendly to the fossil-fuel industry.[11]

A comprehensive review of public opinion literature published in 2017 by Patrick Egan of New York University and Megan Mullin of Duke University showed that unlike many other issues, polarization had not yielded much of a constituency for action on climate change.[12] Even liberals and Democrats who accept climate change science and express concern about global warming's effects ranked the problem well below many other national priorities. With this as a backdrop, Congress has had little incentive to advance major legislation to tackle the problem. The mid-term elections in 2018 brought signs that attitudes are changing, but the jury on long-term, comprehensive action is still out.

With climate change legislation going nowhere in Congress, the Clean Air Act became the primary tool in the Obama administration's limited toolbox to address what many consider to be the most pressing environmental issue of our time. The centerpiece of these efforts, the Clean Power Plan, would become "the Super Bowl" of climate change litigation.[13]

The EPA's role in regulating greenhouse gases began inconspicuously at a 1998 hearing before the House Appropriations Committee. House Majority Whip Tom Delay (R-TX) asked EPA Administrator Carol Browner whether she believed that the Clean Air Act allowed the EPA to regulate emissions of carbon dioxide. Browner responded that she did. A subsequent legal opinion by the EPA's general counsel, Jonathan Cannon, concurred—the Clean Air Act authorizes the EPA to regulate a substance if it is an "air pollutant" and the administrator finds that emissions endanger public health or welfare. Cannon noted, however, that an "endangerment finding" had not yet been made.[14]

When George W. Bush assumed the presidency, the EPA reversed course. The new administration argued that Congress never intended to regulate carbon dioxide and other greenhouse gases under the Clean Air Act. Therefore, the EPA lacked authority to do so. In response, twelve states, three cities, and a host of environmental groups filed suit to force the EPA to regulate greenhouse gas emissions from new motor vehicles. The case, *Massachusetts v. EPA*, became the first U.S. Supreme Court case dealing with climate change and is among the most significant environmental cases to reach the high court. As Jonathan Cannon notes, "*MA v. EPA* does not rise to the level of cultural significance of *Brown v. Board of Education*, but in environmental law, it may prove to be as close as we will come."[15]

As one of the twelve states and others filing suit, Massachusetts got its name on the case because the state convinced the courts that it has a direct stake in the outcome—a requirement in federal law known as "standing." To demonstrate standing, Massachusetts argued that vehicle emissions add to greenhouse gases that contribute to global warming, which in turn endangers its coast.

Meanwhile, industry groups and climate change deniers scorned the idea that carbon dioxide (CO_2) was an air pollutant. The Competitive Enterprise Institute ran television ads that ended with a young girl in a pastoral setting blowing seeds from a dandelion head. With the obvious connection that her breadth contained CO_2, the voiceover said, "Carbon dioxide. Some call it pollution. We call it life."[16]

In 2007, in a huge, but narrow (five to four) victory for environmentalists, the U.S. Supreme Court ruled that the EPA has the authority under the Clean Air Act to regulate CO_2 and other greenhouse gases and that, in declining to regulate, the EPA had improperly failed to make an endangerment finding. The Supreme Court reasoned that the Clean Air Act embraced "all airborne compounds of whatever stripe. . . . Congress may not have had climate change specifically in mind . . . but it gave the EPA authority to address emerging serious problems of precisely this kind."[17] As two Harvard law professors put it, "even the Supreme Court thinks something must be done."[18]

In 2009, with Obama in office, the EPA concluded that "greenhouse gases in the atmosphere may reasonably be anticipated both to endanger public health and to endanger public welfare." The finding was based on substantial scientific documentation. With its

endangerment finding in place, the administration extended its regulatory reach beyond motor vehicles to power plants—a prerogative
upheld by the U.S. Supreme Court in 2011. At the time, power
plants were the largest source of U.S. carbon dioxide emissions,
with aging, coal-fired plants contributing the lion's share.[19]

CONTROLLING POWER PLANT EMISSIONS

The Clean Air Act differs from the Clean Water Act in a fundamental way. While the Clean Water Act applied to both new and existing plants, the Clean Air Act set emission standards only for newly
constructed facilities. This was a major limitation in the Clean Air
Act for coal-fired power plants but, at first, wasn't recognized as
such. A coal plant's useful life was about thirty years. Many had
been operating for decades. The thinking was that it would not be
long until many of the older plants would be replaced by new or
updated plants subject to the emission standards.[20]

This conventional wisdom would prove completely wrong. The
regulations created huge financial incentives to keep the old coal-
fired clunkers running as long as possible. Newly constructed plants
required either investment in multi-million dollar "scrubbers" to
remove pollutants as they pass through the smokestack or use of
low-sulfur coal. Starting in 1978, all new coal-fired plants were
required to install a scrubber, even if they burned low-sulfur coal.
Additional restrictions were placed on emissions of nitrogen oxides.
By 1985, the retirement age for power plants had increased from
thirty years to as long as sixty years.[21]

The Clean Air Act requires existing plants to adopt emission
standards if modifications are made that result in a "significant increase" of a regulated pollutant. "Routine scheduled maintenance"
is exempt. This permitting process is known as a *New Source Review*. What qualified as a modification that triggered New Source
Reviews shifted over time. During the Reagan administration, the
Edison Electric Institute, a trade association, advised utilities to
classify renovations as "upgraded maintenance programs" and any

aspects extending a plant's life as simply "plant restoration projects."[22] The EPA bought into this approach, which allowed plant
upgrades without triggering new emission requirements. The first
Bush administration likewise gave utilities lots of flexibility to mod-
ify plants without kicking in new emission regulations.

It was left to the Clinton administration to begin an aggressive
enforcement campaign. Over one hundred investigators were sent to
more than thirty power plants to comb through records and interview employees. The Department of Justice filed suit against nine
power companies that together controlled well over a third of the
nation's coal-fired generation capacity. Despite the weight of evidence, only one utility (Tampa Electric) settled claims against it
while Clinton was still in office.[23] The others held out for a more
sympathetic White House, which arrived with the election of
George W. Bush.

The Bush administration announced that it was dramatically eas-
ing the New Source Review requirements by administrative action.
As part of a *Safe Harbor Rule*, companies could count any modifications costing less than 20 percent of the total cost of replacing a
unit as "routine maintenance."[24] These lax requirements were dic-
tated by the White House at the behest of utility lobbyists, over the
objections of EPA Administrator Christine Todd Whitman. This
was one of several major disagreements with the White House, and
Whitman resigned in 2003. In 2006, the Safe Harbor Rule was
vacated by the DC Circuit Court, but enforcement under Bush continued as though it was still in effect.[25]

The election of Barack Obama reinvigorated regulating emissions from existing coal-fired power plants in three major ways—
the Cross-State Air Pollution and mercury rules discussed in the
previous chapter, and a focus on greenhouse gases. In 2015, the
EPA finalized the Clean Power Plan, an historic first to reduce
carbon emissions from existing fossil fuel power plants.

The basic concept of the Clean Power Plan was simple. Each
state was given an individual goal for cutting power plant emissions
but could decide for themselves how to get there. They could upgrade their plants, switch from coal to natural gas, expand renew-

able energy, improve energy efficiency, and enact carbon pricing. States would be free to combine any of these options in a flexible manner to meet their targets. They also could join with other states to find the lowest cost options for reducing carbon emissions, including through emissions trading programs. The only part of the Clean Power Plan that wasn't flexible was the timeframe. States had to submit their plans by 2016–2018, start cutting by 2022 at the latest, and then keep cutting through 2030. If states refused to submit a plan, the EPA would impose its own federal plan.[26]

The EPA estimated the plan would reduce power plant emissions of carbon dioxide by about 32 percent by 2030, compared to 2005 levels.[27] This would lead to a several percent cut in overall U.S. greenhouse gas emissions: a good down payment, but not enough to halt global warming. The Clean Power Plan was just one piece of broader needs.

Ramping up renewable energy to reduce emissions was the most controversial part of the Clean Power Plan, as it would rely on actions "outside the fence" of the existing fossil fuel power plants. Industry groups were adamant that the EPA limit itself to the much more modest reductions that could be made "inside the fence" by the power plants themselves—such as substituting fuels or improving the efficiency of furnaces.

The U.S. Chamber of Commerce, coal industry, and lawmakers from coal-based states vehemently opposed the Clean Power Plan. Mike Pence, then governor of Indiana, threatened not to comply. The Competitive Enterprise Institute claimed that it would cause "skyrocketing electric and gas bills" and "threaten to turn out the lights in much of the country."[28] However, not all industry groups were opposed. Utility companies that had a low-carbon fleet of natural gas or nuclear-powered plants were in favor of the plan. The EPA's analysis concluded that for every dollar spent to comply with the Clean Power Plan, the public potentially could get more than six dollars in benefits. These would include climate benefits as well as public health benefits resulting from the accompanying reductions in small particulate air pollution.[29]

Trump's executive order (signed while visiting the EPA) didn't repeal the Clean Power Plan by itself—it simply put everyone on notice of his intents. To repeal or vastly change such a rule, the EPA must go through a formal rule-making process. First, the EPA has to propose a new rule for replacement or repeal, laying out detailed legal and technical justifications for its actions. Next, the EPA must solicit public comments on its proposal. Then, the EPA has to read through all the substantive comments and either take them into account or explain why it's ignoring them. The final rule is subject to judicial review and potential lawsuits, during which the EPA must prove to the courts that its new approach is superior, and that the agency is not acting in an arbitrary or capricious manner.[30]

To appreciate the challenge of this rule-making process, when the Obama administration released its proposed Clean Power Plan in 2014, the EPA consulted with hundreds of groups and received more than four million public comments—by far the most ever for an EPA rule. After working through all these comments, the EPA made numerous changes. When the final rule came out in 2015, industry groups and about half the states challenged it in court. In 2016, the U.S. Supreme Court placed a hold on implementation of the Clean Power Plan until pending lawsuits were resolved. This is where matters stood when Trump issued his executive order to repeal the rule.[31]

Notably, Trump's executive order did not suggest that the EPA would try to revoke the agency's 2009 endangerment finding. Pruitt recognized that given the scientific weight of evidence for human-caused climate change, the legal hurdles to overturn the endangerment finding were overwhelming. By not doing so, however, it was much more difficult to make the case to drastically change or repeal the Clean Power Plan. Many conservative groups were furious at Pruitt. A *Breitbart News* columnist suggested that he resign.[32]

Efforts to replace the Clean Power Plan were a continuation of the Trump administration's fight against any measure to control greenhouse gases—a position brought to worldwide attention when the United States suddenly became a rogue nation by walking away from its commitments to the Paris climate agreement.

On December 12, 2015, representatives of 195 nations reached a landmark accord that committed nearly every country, rich or poor, to take actions that would lower greenhouse gas emissions. After decades of international efforts to address climate change, this was an unprecedented historic breakthrough. Cheers and ovations arose from the thousands of delegates that had gathered in Paris to finalize the agreement. Only two nations among those entitled to be in the pact refused to sign the agreement—Nicaragua (which wanted more financial help for poorer countries) and war-ravaged Syria. Even North Korea ratified the Paris agreement, pledging to wage "war on deforestation" and cut greenhouse gas emissions by 37 percent.[33] In a televised address from the White House, President Obama applauded the deal: "This agreement sends a powerful signal that the world is fully committed to a low-carbon future. . . . We've shown that the world has both the will and the ability to take on this challenge."[34]

By itself, this agreement will not solve global warming. At best, it will cut global greenhouse gas emissions by about half the amount needed to fend off an increase in atmospheric temperatures of two degrees Celsius (3.6 degrees Fahrenheit) above pre-industrial levels. Many scientists consider this temperature increase to be the tipping point beyond which significant harm will occur from rising sea levels, as well as increased severity of droughts, floods, and deadly heat waves.

Countries are required to reconvene every five years, beginning in 2020, with updated plans to tighten their emissions cuts and report on their progress. But there is no legally binding requirement on how much each country should cut emissions. NASA scientist James Hansen, who is credited with first bringing global climate change to national attention, has long criticized the Paris agreement as completely inadequate to address the problem. Nonetheless, the Paris agreement was a major step in the right direction. The hope is that it will create a global system of peer pressure, where no country wants to be viewed as an international laggard.

By allowing countries to develop their own domestic pledges to tackle climate change, Obama was able to sign the agreement by executive order and avoid a ratification battle in the Republican-controlled Senate. But this also made it easier for Trump to change Obama's pledge. In August 2017, the Trump administration rejected the view of modern science on global climate change and announced its intention to withdraw from the Paris agreement. Three months later, Nicaragua President Daniel Ortega said he would sign onto the accord. Syria followed suit the following month, leaving the United States as the only holdout. Ironically, it was the U.S. leadership that paved the way to the Paris agreement; most notably, Obama's Clean Power Plan and joint announcement with Chinese President Xi Jinping that the world's two largest greenhouse gas polluters had agreed to cut their emissions.

The actual reality of the United States withdrawing from the Paris agreement is not so instantaneous or easy as Trump would have liked. Under the terms of the Paris deal, the United States can't withdraw until November 4, 2020—one day after the next presidential election and a month before the EPA turns fifty. Therefore, Trump's notice carried no legal weight, but it did send a strong negative message. When Trump made his announcement, all members of the congressional Republican leadership were united in their praise.[35]

In August 2018, the Trump administration delivered a double whammy to efforts to combat global warming. First, the EPA announced its proposed weakening of auto emission standards. Later that same month, the EPA announced its planned replacement for the Clean Power Plan. Under the proposed replacement, states would create their own rules for *existing* coal-fired power plants based around energy efficiency improvements.[36] The EPA also proposed to ease requirements for New Source Reviews. The end result would eliminate most of the reductions in greenhouse gases envisioned by Obama's Clean Power Plan. A few months later, the EPA proposed to increase the allowable amounts of carbon dioxide emissions from *new* power plants.[37] The proposal would eliminate Obama-era restrictions that, in effect, required newly built coal plants to include carbon capture systems—a technology that is not in use on a commercial scale but viewed by many as an essential future tool for addressing climate change. A coal-fired power plant had not come online in the United States since 2012, but the clear intent was to ease the way for future plants. A push was underway by the EPA to get regulations out the door before the end of 2018, giving the administration two years to defend its environmental policies in court before the end of Trump's current term.

The EPA glibly called the proposed Clean Power Plan replacement the Affordable Clean Energy Rule. The phrase "climate change" was barely mentioned in the nearly three-hundred-page proposal. The EPA even made last minute changes to remove material warning about the dire impacts of climate change.[38] Meanwhile, global temperatures had exceeded the twentieth-century average for more than four hundred consecutive months. The last time the Earth had a cooler-than-average month was when Reagan was elected to his second term.[39]

The Affordable Clean Energy Rule had repercussions for public health as well as climate. The Trump administration's own analysis revealed that, compared to Obama's Clean Power Plan, the new rules could lead to as many as fourteen hundred premature deaths annually from heart and lung disease (due to an increase in fine particulate matter), up to fifteen thousand new cases of upper respiratory problems, a rise in bronchitis, and tens of thousands of missed school days. William Wehrum, head of the EPA's Office of Air and Radiation, called these "collateral effects," but claimed, "We have abundant legal authority to deal with those other pollutants directly, and we have aggressive programs in place that directly target emissions of those pollutants." It was part of a continuing pattern to discount co-benefits of regulations to diminish their perceived value.[40] The EPA finalized the Affordable Clean Energy Rule in July 2019. It was soon challenged on multiple fronts.

Despite every effort of the Trump administration, King Coal is dying. More U.S. coal-fired power plants were shut during Trump's first two years than were retired in the whole of Obama's first term.[41] Changing energy markets favor natural gas and renewables. Gas-fired plants are cheaper and cleaner than coal plants (with caveats on the "cleaner" part that we'll turn to shortly). Tighter air pollution regulations, such as the mercury rule and Cross-State Air Pollution Rule, accelerated the change from coal, forcing utilities to choose between upgrading aging coal-fired power plants to make them cleaner or switching to natural gas or green sources like wind and solar. Many utilities decided to switch. "The history of energy use is a sequence of transitions to sources that are cheaper, cleaner and more flexible," says Vaclav Smil, who has written extensively on energy issues. These transitions are slow, he notes, "but the process is inexorable."[42] The Trump administration cannot stop the transition away from coal, but it can slow it down, hindering progress urgently needed to address climate change.

The energy transition away from coal is an example of how the links between environmental regulation and jobs continues to be played out in bumper-sticker mentality. The health of the U.S. economy depends on broad macroeconomic factors such as the rate of inflation, population growth, and overall demand. Environmental regulations have little to no effect on long-term aggregate employment.[43] Regulations might reduce employment in coal mining but, on the flip side, jobs increase in wind and solar energy. In 2017, twice as many Americans worked in the wind industry as in coal mining. Solar employs many more.[44] Opponents of regulations also focus almost exclusively on economic and employment costs while downplaying the public health benefits, which tend to disproportionately affect lower-income and working-class people who are those most exposed to industrial pollution.

Of course, all this is of little consolation to people who have lost their jobs. Burton Richter, a Nobel laureate in physics, stated the obvious: "You could pension off all the 80,000 workers in the coal industry for a tiny fraction of the medical bills due to burning coal."[45] Not to mention the climate costs. However, it doesn't take a Nobel laureate to make this connection. Taking proactive steps to provide alternate employment opportunities for coal miners is a huge missed opportunity for those who purport to care about these workers.

Many of the benefits of the Clean Power Plan estimated by the Obama EPA were based on calculating, in dollar terms, the economic damages from climate change. Known as the *social cost of carbon*, climate economists call this "the most important number you've never heard of."[46] It applies not only to the cost-benefit analyses used for the Clean Power Plan, but also other regulations on energy efficiency and vehicle fuel economy standards.

In 2009, as a follow-up to the EPA's endangerment finding, the Obama administration created an interagency working group to estimate the social cost of carbon. This is a "fiendishly difficult task," noted *The Economist* magazine, without much exaggeration. The calculation required not only modeling climate change but also its impact on human health and economic productivity. The working group estimated that the social cost of a ton of carbon dioxide was about forty-seven dollars.[47] That is, for every ton of carbon dioxide emitted, the costs of the climate impact—such as coastal erosion, flooding, reduced agricultural outputs, and increased disease—are estimated to add up to forty-seven dollars. This estimate obviously has huge error bars, but the approach is central in trying to determine a ballpark indication of how much society should pay now to avoid future pain. Many economists consider forty-seven dollars per ton too low, as it leaves out many impacts that are unknown or difficult to value.

In March 2017, Trump signed an executive order disbanding the working group and withdrawing all its reports and findings. The Trump administration knew that eliminating the social cost of carbon would have a tough time in court, so instead, decided to lowball it, dropping the calculated social cost of carbon to around one to seven dollars—a reduction of 97 to 87 percent.[48] This significantly diminished the calculated economic benefits of regulating CO_2.

The administration achieved its reduction in the social cost of carbon in two major ways. The first was to use a "discount rate" that gives less weight to the future. The second was to include only the *direct* benefits of reducing greenhouse gas emissions within the U.S. borders. Excluding effects outside the United States overlooks two major issues. Global migration, economic destabilization, and political unrest caused by climate change in other countries will, one way or another, affect the United States.[49] Second, climate change can only be meaningfully addressed by all countries working together. "The United States is only 14 percent of global emissions, which means that 86 percent of the damages we face will be caused by emissions from other countries," said Richard G. Newell, president of Resources for the Future, an environmental economics think tank. "If we took this domestic-only approach to its logical conclusion, that means other countries should not worry about their impacts on us."[50]

On a final note, the Clean Power Plan would not be anyone's first choice for regulating carbon. It was purely a fallback position by Obama to address climate change "with or without Congress."[51] A tax on carbon emissions that would increase every year until emissions goals are met would be a much more efficient approach. The tax could be rebated equally to consumers and/or used to reduce income taxes. The tax would send a price signal that replaces the need for less efficient carbon regulations and promotes investment in clean energy technology and energy conservation. In January 2019, forty-five prominent economists published a statement in support of a carbon tax in the *Wall Street Journal*. Within a few months the declaration had been signed by thirty-five hundred economists, including twenty-seven Nobel laureates.[52] A carbon tax also has the support of a broad array of prominent Republicans and businesses like BP, Royal Dutch Shell, Pepsi, and General Motors.[53] But this approach is politically untenable as long as climate change deniers are in key positions.

Also downplayed in the climate debates is the cost to the United States in terms of future competitiveness. As noted by former EPA Administrator Gina McCarthy: "From catalytic converters to smoke-stack scrubbers, America has a legacy of innovating the world's leading environmental technologies—accounting for more than 1.5 million jobs and $44 billion in exports in 2008 alone."[54] In the Trump administration, these facts fall on deaf ears.

Trump often talks about how China is "killing us" on the economic front. Directly beneath the front-page article in the *New York Times* covering the EPA's repeal of the Clean Power Plan was an article by CNN's Fareed Zakaria detailing China's massive investment in electric vehicles as part of its determination to dominate clean energy technology. While China pursues these goals, the Trump administration is "engaged in a futile and quixotic quest to revive the industries of the past," said Zakaria. "Who do you think will win?"[55]

METHANE

While carbon dioxide is the most notorious greenhouse gas, methane holds the number two slot. As the key constituent of natural gas, it's not surprising that the oil and gas industry is a major methane source. Emissions of methane come from flaring gas during oil production and from leaks around natural gas wells, storage tanks, refineries, and pipelines. Other methane sources include landfills and animal feeding operations. Methane emissions also arise naturally from wetlands.

Methane is much more powerful than CO_2 at absorbing radiation. The good news is that it's much shorter lived. While CO_2 persists in the atmosphere for centuries (or even millennia), methane warms the planet for a decade or two before decaying to carbon dioxide and water vapor. The bad news is that methane's global warming potential is eighty-six times that of carbon dioxide over twenty years, and thirty-four times greater over one hundred years. Any way you look at it, methane is a potent greenhouse gas.[56]

In 2014, Colorado enacted the nation's first comprehensive regulations to directly control methane (and volatile organic compound) emissions from oil and gas operations, requiring energy companies to regularly inspect oil field equipment for leaks. The rules were endorsed by Colorado Governor John Hickenlooper, a former petroleum company geologist, and supported by the Environmental Defense Fund and three of the state's largest oil and gas producers.

The next year, Obama made a national commitment to reduce methane emissions from oil and gas operations by 40 to 45 percent from 2012 levels by 2025. In May 2016, the EPA announced the first-ever federal rules targeting emissions from *new or modified* oil and gas operations.[57] Although this was a good first step, regulating new sources would only achieve part of the reductions needed to meet Obama's goals. The question remained: What to do about the hundreds of thousands of *existing* oil and gas sources that remained unregulated? To help develop such regulations, the EPA sought broad-based information from oil and gas companies. But the timing couldn't have been worse. The finalized information request was made two days after Trump's election.

Another effort to reduce methane emissions also came days after the election, when Secretary of the Interior Sally Jewell announced the methane Waste Prevention Rule, which restricts flaring of methane at oil wells on public and tribal lands. Operators must also periodically inspect their operations for leaks and replace outdated equipment that vents large quantities of gas into the air. An estimated $330 million a year in methane is wasted through leaks or intentional releases on federal lands. Senator Tom Udall (D-NM) noted that the Waste Prevention Rule would provide badly needed revenue to states like New Mexico. The more methane that is captured on federal lands, the more money that flows into government coffers. "This rule is simply good policy—good for taxpayers, good for the economy, and good for the environment," Udall said.[58]

Reducing methane emissions from oil and gas operations is an obvious step to make natural gas more of the "clean fuel" claimed by its proponents. Yet industry balked at all three of these actions. "Methane is the product that we sell. We are incentivized already to prevent methane emissions," claimed Howard Feldman, senior director of regulatory and scientific affairs for the American Petroleum Institute.[59] Ironically, in many ways the oil and gas industry fared well under Obama. His administration was supportive of the controversial practice of hydraulic fracturing for unconventional oil and gas resources and also lifted the forty-year-old ban on oil exports, a considerable potential boost to profits.[60]

Obama's efforts to reduce methane emissions from oil and gas operations were an early target of the Trump administration, but with mixed success. Pruitt withdrew the information request on existing operations with the stroke of a pen, indicating a lack of interest in reducing emissions from these sources. On the other hand, the rules targeting emissions from *new or modified* operations had been in place long enough that they could only be changed by a laborious rule-making process. Pruitt tried to delay their implementation by two years but was turned back by the courts.[61]

In May 2017, Republicans sought to repeal the Interior Department's Waste Prevention Rule using an obscure law known as the Congressional Review Act (CRA). This law, the brainchild of Newt Gingrich in the 1990s, gives lawmakers sixty days to review new rules. During this period, they can overturn an agency regulation with simple majority votes in both the House and Senate. If a regulation is nullified, a similar rule cannot be reissued without explicit approval from Congress, even under a future president. This measure had been used only once—in 2001 to overturn a rule on workplace ergonomics.[62]

In a flurry of regulatory rollback, Republicans used the CRA to eliminate fourteen regulations that had been issued during final months of the Obama administration. This approach failed for the Interior Department's methane Waste Prevention Rule, however, as three Republican Senators voted against the repeal. These Senators didn't want to preclude the option of replacing the methane rule with alternate regulations. With the CRA option off the table, Trump's Interior Secretary Ryan Zinke delayed enforcement of the methane rule and initiated a formal rule-making process to permanently rewrite or undo it.[63]

Trump's EPA continued its efforts to neuter the Obama-era methane regulations. In September 2018, the EPA announced it would be undertaking "targeted improvements" to the Obama administration's rule on *new or modified* oil and gas operations by reducing requirements to monitor and repair leaks. This proposal would save industry up to seventy-five million dollars in regulatory costs annually.[64] In 2019, the EPA went one step further, proposing to eliminate federal requirements to monitor and repair methane leaks. Meanwhile, a recent study suggests global methane emissions may have been underestimated.[65]

We have focused on oil and gas operations, but it should be noted that agriculture is another major source of methane. About a quarter of U.S. methane emissions come directly from livestock, much of it from belching. Manure lagoons generate about a tenth of all methane emissions. Algal blooms from nutrients in waterways are another source. Despite the potential impacts of climate change on agriculture, farmers and ranchers tend to be against any sort of climate-based regulations.

HYDROFLUOROCARBONS

A third greenhouse gas of concern, hydrofluorocarbons (HFCs), is a classic case of trying to solve one environmental problem and creating another. It's also a case of strange bedfellows, with Obama's EPA, large U.S. chemical companies, the U.S. Chamber of Congress, and environmentalists on the same side. To understand why, we have to go back to concerns about a class of chemicals known as chlorofluorocarbons (CFCs). Widely used in refrigeration, air conditioning, aerosol sprays, and foam-blowing agents, studies revealed that CFCs were destroying the stratospheric ozone layer that shields us from harmful ultraviolet rays.

Despite the dangers, banning CFCs became highly controversial. During the Reagan administration, Interior Secretary Donald P. Hodel famously suggested that instead of curbing CFCs, people should combat the added risk of skin cancer by protecting themselves with hats, long-sleeved shirts, and sunscreen.[66] The scientific evidence for banning CFCs was fought intensely by many of the same people who also denied the overwhelming scientific evidence linking smoking to lung cancer and power plant emissions to acid rain. Some of these same people and groups are now at the center of debunking human-induced climate change.

The 1987 Montreal Protocol, the international treaty adopted to restore the Earth's protective ozone layer, is one of the great environmental success stories of modern times. It was universally ratified by all countries. With its focus on the ozone layer, climate change was not considered during treaty negotiations in the 1980s. Yet the agreement resulted in a side benefit of reducing greenhouse gas emissions. CFCs not only deplete the ozone layer; they are also powerful greenhouse gases.

After CFCs were banned, use of HFCs grew rapidly as a substitute. But there was a major problem. Whereas use of HFCs may be helpful in protecting the ozone layer, they are also extremely potent greenhouse gases, with an Earth-warming potential of hundreds to thousands of times that of carbon dioxide. Fortunately, they're released in much smaller quantities. They also break down in as little as fifteen years, so reductions in the use of HFCs could yield quick results. As part of its commitment to reduce greenhouse gases, the EPA under the Obama administration issued a rule in 2015 restricting the use of HFCs.

Two foreign-owned companies that produce HFCs—Mexichem and Arkema—sued to keep these chemicals on the market. The legal question at the center of the lawsuit was whether the EPA could use a section of the Clean Air Act geared toward phasing out ozone-depleting substances to replace HFCs. In August 2017, a federal court ruled that the EPA could not stop companies already using HFCs from continuing to do so. The opinion was written by Judge Brett Kavanaugh and provides a window into his likely future positions on environmental cases that reach the Supreme Court. During his time in the DC Circuit court, Kavanaugh granted federal agencies power to craft rules only when Congress clearly spelled out in the law that that is exactly what it wanted. In the HFC case, he wrote: "However much we might sympathize or agree with EPA's policy objectives, EPA may act only within the boundaries of its statutory authority. Here, EPA exceeded that authority."[67]

The Natural Resources Defense Council, Honeywell, and Chemours (a spin-off of DuPont) appealed the judges' decision, arguing that the EPA should be entitled to more leeway in interpreting the Clean Air Act. Honeywell and Chemours had invested more than a billion dollars to make alternative refrigerants to replace the HFCs. In October 2018, the Supreme Court turned down the appeal, the day after Kavanaugh was sworn into the high court.[68] Separately, the EPA had announced the previous month that regulations to prevent leaking of ozone-depleting refrigerants during repair, maintenance, and disposal of appliances would not apply to HFCs.

The climatic effects of HFCs achieved international attention through an amendment to the Montreal Protocol. Adopted in 2016, the Kigali Amendment is named for the capital of Rwanda where it was finalized. Countries that ratify the Kigali Amendment commit to cut the production and consumption of HFCs by more than 80 percent over the next thirty years. The HFC phasedown is expected to avoid up to 0.5 degree Celsius of global temperature rise by the end of the century, while continuing to protect the ozone layer. U.S. manufacturers say the agreement could create tens of thousands of jobs and generate billions of dollars in U.S. exports, while also helping to protect the planet. Despite broad industry support and a letter from thirteen GOP senators urging him to send the treaty to the Senate for ratification, Trump failed to do so.[69] When the Kigali Amendment went into force at the beginning of 2019, it was ratified by sixty-five nations. The United States was not among them.

The Trump administration has shown outright hostility toward any efforts to combat climate change, including those supported by U.S. businesses. The administration's actions will prolong the life of dirty coal-fired power plants, increase tailpipe emissions of greenhouse gases from cars and trucks, and allow oil and gas operations to continue spewing large amounts of methane into the atmosphere. Trump's disdain for action on climate change is so extreme that after announcing the U.S. planned withdrawal from the Paris climate agreement, he sent a representative to follow-on G7 discussions on global warming for the sole purpose to promote fossil fuels.[70]

With the clock ticking, the Trump administration continues to scorn climate science. In 2018, when the most recent UN Intergovernmental Panel on Climate Change report came out, it argued that world leaders' pledge to keep warming below two degrees Celsius is too modest, and that they have about a decade to get on track. The report was written by ninety-one leading scientists from forty countries who together examined more than six thousand scientific studies. With his usual backhanded way of casting doubt, Trump's response was "who drew it. . . . I can give you reports that are fabulous and I can give you reports that aren't so good."[71]

Part IV

Toxic Chemicals and Hazardous Waste

9

TOXIC CHEMICALS

We learn geology the morning after the earthquake.
—Ralph Waldo Emerson

When President Obama signed the reformed Toxic Substances Control Act (TSCA) on June 22, 2016, this environmental milestone was only marginally covered in the news. Captivated by Donald Trump's campaign for president, the media had flocked to the Trump SoHo Hotel in New York City to cover his speech about how his opponent was a criminal and should be locked up. Then he threw in a new one—on top of everything else, Hillary Clinton was "a world-class liar." The media just couldn't get enough of this stuff. Meanwhile, what the president had just signed back at the White House was actual news; nothing short of a miracle, really. After years of debate and inaction, Congress finally had taken bilateral action to reform the all-but-toothless TSCA.

It's almost instinctive to bash chemical companies, yet many chemicals make it possible for us to live longer, more comfortably, and safely. Exposure to chemicals is a price we pay for the conveniences of modern life. No chemical is totally innocuous, but some come with a sufficient downside that they should be restricted in their use. Others are so bad that, no matter what benefits they may bestow, you just can't have them around. As with truly horrific crimes that, for many, justify the death penalty, there's also a death penalty for the really bad chemicals—and this is one of the most complicated jobs that the U.S. Environmental Protection Agency (EPA) must tackle. Like its human counterpart, sentencing a chemical to death is expensive and time-consuming just to prepare the case, and then there's the trial. Or, more accurately, trials.

Controversies often involve not only *what* chemicals, but also *which uses* of a chemical to regulate. Complicating matters, the effects of exposure to a hazardous substance depend on numerous factors—the dose, the duration, how you are exposed (breathing, eating, drinking, or skin contact), personal traits and habits (age, sex, diet, family traits, lifestyle, and general state of health), and whether other chemicals are present. [1]

BEFORE THE EPA

In the early years, the American chemical industry was a small group of disparate fiefdoms each basically going its own way. Then along came World War I, also known as "The Chemists' War." When the war began, Germany dominated the chemical industry. It wasn't long before wartime blockades and shortages created the need for new chemical suppliers and new chemicals. To meet the requirements of Allied war contracts, firms like Dow Chemical Company and DuPont began researching a variety of raw materials and developing alternatives. Then, in the spring of 1915, the trenches on the western front descended into unmitigated hell with the introduction of gas warfare. U.S. chemical companies began churning out an arsenal of chloropicrin, phosgene, and mustard gas. [2]

The production capacity required to wage war on the world scale created a new partnership between the federal government and chemical firms. By war's end, the American chemical industry had come of age as a politically powerful confederation characterized by intensive research, rapid product development, and enormous capabilities of production. The industry continued to build on its wartime research and production while diversifying into peacetime chemical applications, such as pesticides and pharmaceuticals.

The industry had cozied up with the government for the war effort and reaped massive rewards because of it, but there was nothing new about their post-war modus operandi. From its earliest days, the American chemical industry had virtual carte blanche to do exactly as it pleased, including the freedom to foul its nest and everyone else's without meddling interference from government.

Teddy Roosevelt was an early exception. Upon becoming president in 1901, he took on Wall Street and managed to "bust the trusts." Roosevelt was also the first environmental president. He cared about our priceless and irreplaceable natural heritage and set aside vast chunks of land to be preserved as National Parks, Forests, Grasslands, and Monuments.

While governor of New York, Teddy kept a wrestling mat at his office. Male visitors were invited to strip down to their skivvies and have a good rough and tumble with him. What better way to get the real measure of friend or foe, or (as these things go) *both*. It was a novelty, a real ice-breaker. It also let everyone know that Teddy loved a good fight. And he loved to win.

As president, one of Roosevelt's biggest fights was with Anaconda's copper mining and smelter operation in Montana—the largest in the world. Anaconda had originally built its smelter near its mining operation in Butte. After repeated episodes of deaths caused by rampant air pollution, Anaconda moved the smelter to a remote valley. It wasn't long before crops and livestock were dying, along with trees in the national forest. An old photograph from the Montana Historical Society captures an eerie resemblance to Dante's purported lowest level of Hell, as the smelter's smokestacks spew smoke over hill and dale in such massive doses that nearby mountains were more or less erased. This nonstop cloud of emissions contained high levels of arsenic and other toxic metals.

The valley's ranchers were furious. Roosevelt, having once been a cattle rancher in the Dakotas, *got it*. He tried the diplomatic approach, inviting the company chieftains to the White House. But all attempts at compromise were rebuffed. As a subsidiary of Standard

Oil, Anaconda could count on Wall Street's support. Roosevelt's attorney general laid out the difficulties: "The mouths of all the experts would be closed because [Anaconda] and the Standard Oil Company ultimately will reach them and control them."[3] Nonetheless, in one of the first environmental confrontations of industry by government, Roosevelt initiated a lawsuit to force Anaconda to curb its air emissions.

Fightin' Teddy lost, and in a big way. Under the eventual settlement agreement, industry was revealed in all its power. The government not only had to drop its insistence on the company controlling emissions but had to accept in its place Anaconda's criterion for action—emission controls would be employed *only* when the captured emissions could be sold for a profit.[4]

As it turned out, arsenic not only could be sold, but was soon a big moneymaker. Market demand grew rapidly after it was discovered that calcium arsenate could control that curse of cotton crops— the infamous boll weevil. Another breakthrough was use of lead-arsenic pesticides for controlling moth infestations in fruit orchards. Moths are big fans of fruit orchards. Their specially adapted tongues allow them to pierce into fruit and gorge on the juice. Spraying fruit trees with a lead-arsenic pesticide was a highly effective deterrent, and it soon came into wide use.

In 1919, a Boston health inspector discovered fruit flecked with white spots. Laboratory tests determined that the spots were arsenic compounds. Fruit with high levels of arsenic was soon seized by health authorities in Boston, Los Angeles, and other cities. But it was a futile gesture. All that federal regulators could do was to try to persuade the chemical industry to stop producing this pesticide and the powerful farm lobby to stop spraying it on food. This policy of persuasion had virtually no effect. The use of arsenic pesticides continued to grow.[5]

Other countries were not so lax in protecting the public. In Great Britain, strict standards for arsenic in food had been in place since the early twentieth century, when contaminated beer caused a severe outbreak of arsenic poisoning. A Royal Commission, led by the renowned physicist Lord Kelvin, found that arsenic could be reliably detected at very low levels. The commission recommended a goal of no detectable arsenic in food—translating into an enforceable standard of one part per million. The premise underlying Lord Kelvin's recommendation has become known as the *precautionary principle*. The commission explained, "in the absence of fuller knowledge than is at present available . . . we are not prepared to allow that it would be right to declare any quantity of arsenic, however small, as admissible in beer or in any food."[6] In other words, better safe than sorry. Most of Europe adopted the Royal Commission's arsenic standard as an upper limit for pesticide residue on fruits and vegetables. In the United States, arsenic spraying continued unabated.

In 1925, after an English family was poisoned by arsenic, widespread inspections of American imported fruit to Britain revealed high levels of arsenic on apples. When British authorities threatened to ban fruit imports from America, the U.S. Department of Agriculture began inspecting apples destined for export. Any fruit exceeding Europe's tolerance standard was embargoed and sold domestically. The creation of this system was not announced to the American public for fear of hurting domestic apple sales.

The Department of Agriculture was caught in a bind with a double standard of allowing Americans to eat poisoned apples deemed unsafe for Europeans. The U.S. Food and Drug Administration (FDA) upped the ante when it began studying human tolerance levels of pesticide residues. By feeding low levels of lead arsenate to rats over a lifetime, the FDA sought a scientific basis for setting and enforcing tolerances. This study came to the attention of Representative Clarence Cannon of Missouri, an apple grower who thought residue regulation was "nonsense." Cannon became chairman of the subcommittee that funded the FDA. Congress soon acted by forbidding the FDA to conduct research on the health effects of pesticides, and while they were at it, cut fifty thousand dollars from the agency's annual budget. This was a substantial sum at a time when the government's entire cancer research budget was $115,000 a year.

Cannon was not finished. If lead and arsenic were to be openly allowed in food, they needed their own scientific validation. For this, he brought in Royd Sayers, who had led the investigation of coal-mining diseases for the state of Pennsylvania and played a major role in minimizing the dangers of coal mine dust. Also brought on board was Felix Wormser, secretary of the Lead Industries Association. Wormser had used his control of industry research funding to direct scientists away from the dangers of lead paint on toys and cribs.

To bias the results in industry's favor, former employees (as well as any current workers) who displayed symptoms of pesticide poisoning were excluded from the study. Even then, adverse findings could not be avoided. To solve this problem, they buried the information deep in the report, in a section listing cases that didn't "come up to the study's criteria." The industry report, sent to the FDA in June 1940, recommended more than a *doubling* of the arsenic tolerance, and a *tripling* of the lead tolerance. One of the report's conclusions even argued that combining arsenic with lead might make the lead less poisonous. In August 1940, the arsenic and lead tolerances were officially increased.

The lead-arsenic controversy is just one example of what happens when you have unfettered industry with no regulatory oversight. Over the next three decades, the American chemical industry continued manufacturing new chemicals at a breakneck pace. By the 1970s, tens of thousands of chemicals were on the market, most of which had not been studied for public health or environmental safety. The government's limited budgets for studying the growing problems of chemical pollution perpetuated an unfair playing field. Chemical companies insisted they were the most qualified to study any problems in their own labs, at their own pace. Research of this kind, however, posed a huge risk—a chemical found to be highly toxic or carcinogenic could trigger the demand for control. Therefore, the first line of defense was simply not to investigate. The industry maintained that chemicals were innocent until proven guilty, therefore lack of knowledge justified lack of regulation. When a study could not be avoided (as in the case of lead-arsenic pesticides, leaded gasoline, and black lung disease), "friendly" researchers would slant experiments and cherry-pick data in order to reassure the public that everything was under control. [7]

Chemical contamination continued unabated, but the problem generally received little public attention outside of the affected communities until 1962. That year, Rachel Carson's blockbuster book *Silent Spring* brought national attention to the environmental harm caused by DDT and other pesticides that move up through the food chain. The creation of the EPA brought toxic chemicals further into the spotlight, and contamination events were increasingly gaining national attention.

In 1975, factory workers at a Virginia chemical company, ironically called Life Sciences Products, came down with severe nervous system disorders from the company's product, a pesticide known as Kepone. Twenty-nine factory workers were hospitalized, and the Virginia governor shut down fishing on the James River due to concerns about discharges from the factory. The event attracted national media attention, including an episode by Dan Rather on *60 Minutes*. Although the factory operated for only sixteen months, the event greatly increased public awareness about the potential dangers of chemicals. [8] Another high-profile contaminant, polychlorinated biphenyls (PCBs for short), caused much longer lasting concerns.

PCBs had been in use since the late 1920s, but they weren't on the public's radar until a Swedish research chemist stumbled onto them by accident. In 1964, Dr. Soren Jensen was studying DDT levels in human blood when a mysterious group of chemical compounds kept showing up in his samples and interfering with his analyses. This mystery chemical was turning up everywhere he looked—in fish, sea birds, and human hair. He found it in a sample of his hair. He tested his wife's and children's hair, and there it was. The highest concentrations among his three children were in his nursing infant daughter's hair. As he kept sampling, it appeared that all of Sweden and its adjacent seas had anywhere from trace amounts to much higher concentrations of this chemical compound. [9] But what was it? And more importantly, was it dangerous?

A good research chemist must be part detective, so Jensen began to eliminate suspects. Although chemically similar to DDT, he knew it wasn't a pesticide because he found it in wildlife specimens collected years before chlorine-based pesticides were in general use. An important clue came as he began to study Sweden's archive of eagle feathers dating back to 1888. The mystery compound didn't show up until 1942, then soon began increasing. Jensen eventually identified the chemical structure and became convinced that he was dealing with chlorinated biphenyls, but he still didn't have the faintest idea how they were being used or where they were coming from. Finally, a German chemical manufacturer provided Jensen with a sample of PCBs. It matched his chemical readings.[10]

In 1966, Jensen published an article in *New Scientist* where he laid out his findings of PBC contamination in Sweden, London, and Hamburg. He concluded that PCBs can "therefore be presumed to be widespread throughout the world."[11] This got readers' attention. A key part of the problem is that PCBs bioaccumulate, working their way up through the food chain. When they reach the top of the food chain—humans, whales, polar bears, and dolphins—they are more or less permanently stored at concentrated levels. Jensen's *New Scientist* article caused a stir, but the dangers were still unknown.

The first well-publicized warning that PCBs are highly toxic came just two years later, in 1968, when thirteen hundred people in Japan were poisoned by rice-bran oil (yusho) that had become contaminated with PCBs. Symptoms included severe skin eruptions, discoloration of the lips and nails, and swelling of the joints. Some of the women were pregnant. Of the eleven babies born to these mothers, two were stillborn. Follow-up studies found that the surviving babies exhibited delayed growth, low IQs, and a generally apathetic and dull demeanor. Studies conducted a decade later of adult victims found that the rate of liver cancer was fifteen times higher among the victims than in the normal population.[12] This tragedy changed the meaning of *yusho* in the Japanese lexicon. What had been simply the name for rice-bran oil is now the word memorializing this mass PCB poisoning—*Yusho*.[13]

In 1968, Robert Risebrough of the University of California extended the concerns about PCBs to the United States, when he and his colleagues reported a link between exposure of California's peregrine falcons to PCBs and their failure to reproduce.[14] In 1970, the population in the state was listed at just five pairs (it's since made a remarkable recovery). By the early 1970s, it was well established that PCBs had contaminated the United States, Canada, and Europe. The Arctic and other very remote parts of the world were also contaminated. Asia and Latin America hadn't been systematically studied but were assumed to be contaminated. On top of all this, there was serious alarm about the persistence of these compounds. As Jensen had discovered a few years earlier, the darn things just didn't break down.

But there was still the unanswered question—are PCBs dangerous at environmental levels? The Yusho poisoning in Japan seemed to answer this question once and for all, but the problem was that this was an industrial accident that had resulted in a large release of PCBs directly into food. The more important question remained unanswered—were they dangerous at levels showing up in people and marine life worldwide? Scientists began studying this question, but studies take time—especially longitudinal studies to determine if something is carcinogenic. Nonetheless, the evidence continued to grow.

In the 1970s, three seal species in the Baltic Sea were in decline due to 80 percent of the females being infertile. Was this because of DDT, PCBs, or both? Over time, studies examining damage to wildlife determined that PCBs were the problem. In 1988, the journal *Environmental Pollution* published an article revealing that marine mammals at the top of the food chain (dolphins, porpoises, and whales) all had levels of PCBs far exceeding their terrestrial counterparts—and at seventeen times the level that required manufactured goods containing PCBs to be labeled and handled as toxic waste. The article concluded that marine mammals are acutely sensitive to PCB hormonal effects and may be threatened with extinction.[15]

Evidence was also accumulating that PCBs could affect the unborn child. A study of four-year-old children whose mothers' diets during their pregnancies had included significant amounts of fish from Lake Michigan discovered *all* the children had low birth weight and cognitive deficits. Over time, PCB contamination has been linked to various human health issues, including liver disease, immune system damage, endocrine system disruption, and learning disabilities in children. PCBs are a proven carcinogen in test animals and considered a probable human carcinogen. [16]

In the midst of all these immense challenges to understand the spread of PCB contamination and threats to humans and wildlife, the task of assigning responsibility couldn't have been easier. This wasn't one of those John Travolta, *A Civil Action*, legal conundrums.

Only one company made PCBs, and that company was Monsanto.

When Monsanto began manufacturing PCBs in 1935, it wasn't long before they knew they had a winner. With electricity coming into widespread use during the first half of the twentieth century, PCBs became one of the chemical heroes of the modern age. As electrical wires were run across the country—lighting up towns and cities, ramping up industrial production, and freeing Americans from the never-ending labor of hauling water and chopping wood— the benefits of this miraculous electricity came with the dangers of fire. PCBs were simply made to order as the perfect insulator for electrical transformers and capacitors, and for making fire-resistant coatings on electrical wire and electronic components. A Monsanto engineer called PCBs "as perfect as any industrial chemical can be."

The possibilities seemed endless. It wasn't long before PCBs were being used as plasticizers in paints and cements, in pesticides and oils, in clothing flame retardants, in caulking and sealants, in adhesives and wood floor finishes, in waterproofing compounds, in carbonless copy paper, in dishwasher detergent, and eventually even in surgical implants. Monsanto had hit the jackpot. In addition to their two production plants in the United States, they licensed PCB plants in Austria, France, Great Britain, Italy, Japan, Spain, and what were then the Federal Republic of Germany and the USSR. Production quotas soared, and money kept rolling in.

From a legal standpoint, there were two questions: In the early years, did Monsanto know about the environmental contamination, and did they know PCBs were dangerous? The answer to the first question (except for the immediate vicinity around their plants) is probably no. The answer to the second question is yes.

The toxicity associated with PCBs was recognized very early. In 1937, the Harvard School of Public Health hosted a meeting on the health effects of PCBs after three workers exposed to PCBs at Halowax Corporation had died and autopsies of two revealed severe liver damage. Chloracne (acne-like skin eruptions) also was prevalent among the workers, with some in "very bad condition." Representatives from Monsanto and several other chemical companies attended. The Harvard University researcher reported that his test rats exposed to PCB vapors similar to the workers' conditions also had suffered severe liver damage. The meeting's notes recorded that PCBs are "capable of doing harm in very low concentrations" and are probably the most dangerous of the chlorinated hydrocarbons. The meeting concluded by the Halowax president stressing the "necessity of not creating mob hysteria on the part of workmen in the plants."[17] The results were published but did not gain much attention.

In 1966, Soren Jensen's report suddenly catapulted PCBs from a problem within Monsanto plants to the global environment. It wasn't long before the lawsuits were flying, and Monsanto was court-ordered to turn over pertinent internal documents held by a law firm representing the company.[18] What these documents make clear is that the company's executives cared about two things: dodging lawsuits and protecting profits. Nowhere in all this evidence was there an indication that Monsanto executives felt the least concern, or remorse, for how their products had contaminated most of the planet, and the consequences thereof. Nowhere was there a hint of restricting or shutting down production.

PCB production in the United States continued to grow, peaking in 1970, at eighty-five million pounds.[19] As the facts came out in court, it became obvious that Monsanto had concealed, misrepresented, downplayed, and manipulated evidence of toxicity and environmental harm in order to protect profits. A company document from 1970 basically says it all: "We can't afford to lose one dollar of business. Our attitude in discussing this subject with our customer will be the deciding factor in our success or failure in retaining all our present business. Good luck."[20]

The legacy of PCBs continues to this day. They remain the leading cause of impairment of lakes and reservoirs in five states.[21] PCBs continue to be linked to the decline of marine species such as killer whales.[22] They're also a recalcitrant contaminant at many Superfund sites. The most infamous is a two-hundred-mile stretch of the Hudson River, where two General Electric (GE) capacitor manufacturing plants discharged about 1.3 million pounds of PCBs over a thirty-year period. The PCBs mixed with sediments on the river bottom and collected in fish tissue, prompting health advisories and devastating a commercial fishing industry that had existed for more than a century. After battling the EPA for two decades, GE finally agreed to a massive dredging project to remediate PCB-contaminated sediment "hot spots" along a forty-mile stretch of the Upper Hudson. The battles continue over how much dredging is enough.[23]

All this leads us to one final question. Was Monsanto some kind of outlier? Did it do anything many other chemical companies wouldn't do to protect their products and profits? The answer is a simple, and alarming, no.

THE TOXIC SUBSTANCES CONTROL ACT

The 1976 TSCA was intended to be one of the nation's foundational environmental laws. Most people have never heard of it. When the EPA began implementing the TSCA, there were around sixty-two thousand chemicals (called "existing chemicals" in the law). Since then, about six hundred "new chemicals" have been added each year.[24] These numbers don't include drugs, cosmetics, food additives, and pesticides—the first three of which are regulated by the FDA. The EPA shares responsibility with the FDA for regulating pesticides. (We focus on chemicals under the TSCA but note that regulating pesticides is also one of the EPA's most challenging and controversial tasks.)

The TSCA gave the EPA the authority to regulate toxic chemicals (and ban the really bad ones) but made it almost impossible for them to actually do so. For starters, the TSCA allowed all sixty-two thousand existing chemicals to be grandfathered in as "safe." Second, if the EPA later discovered that one of these chemicals posed an "unreasonable risk" to human health or the environment, the burden of proof was on the EPA.[25]

Twenty years after the act had been passed, the EPA started a voluntary program to try to get chemical companies on board. Under this program, companies were asked to "sponsor" chemicals produced or imported in amounts exceeding a million pounds a year. The thinking was that chemicals in greatest use potentially posed the greatest risk. In other words, if one of them was a problem, then it was going to be a big problem. "Sponsor" was a very elastic term, meaning anything from conducting additional testing to just providing data from existing company files. This effort had some marginal success, with companies sponsoring more than twenty-two hundred chemicals, but it generated only very basic screening-level data for the EPA.[26] The program became largely a public relations opportunity for chemical companies to show they were on board with the EPA.

By the time of the TSCA reforms in 2016 (forty years after passage of the act), the EPA had banned uses of only five of the sixty-two thousand "existing chemicals" under the act—PCBs, asbestos, dioxin, chromium-6, and fully halogenated chlorofluoroalkanes. PCBs were probably the biggest driver behind the original TSCA being passed. They were also banned as part of the act, making PCBs the only contaminant ever directly banned by Congress. The growing recognition of stratospheric ozone depletion also had been among the driving forces for passing the TSCA. To protect the ozone layer, the EPA banned the use of fully halogenated chlorofluoroalkanes (a group of compounds that includes CFCs) in most aerosol spray containers in 1978. A decade later, further regulations on air conditioners and other CFC uses were imposed through the Montreal Protocol. The other three chemicals (asbestos, dioxin, and chromium-6) were also slam-dunks for regulation.

Asbestos started out like PCBs—you just couldn't say enough good about it. Considered a "miracle mineral" due to its strength, durability, and resistance to heat, fire, and electricity, asbestos was widely used in building materials and fireproofing in schools and homes. Today, asbestos is a known human carcinogen linked to lung cancer and mesothelioma. It also causes asbestosis, an emphysema-like lung disease that is disabling and often fatal. Lung cancer has many causes, but most cases of mesothelioma can be linked to exposure to asbestos by workers, those living near asbestos factories, or even family members exposed to workers coming home with dust on their clothes. Mesothelioma typically takes several decades to develop after exposure, which explains all those law firms running television commercials to seek victims for lawsuits. Asbestos is probably the most litigated environmental health concern known to man. Trump, the owner of buildings with asbestos, claims that asbestos is "100 percent safe, once applied" and that "the movement against asbestos was led by the mob, because it was often mob-related companies that would do the asbestos removal."[27]

Dioxins are highly toxic and can cause cancer, reproductive and developmental problems, damage to the immune system, and hormone interference.[28] Chemically similar to PCBs, dioxins are persistent and accumulate in the food chain, mainly via the fatty tissue of animals. This class of chemicals was frequently highlighted in the media as "the most toxic man-made chemical." Unlike PCBs and asbestos, dioxins were never intentionally produced as marketable products, but rather arise as unwanted byproducts. Dioxins were brought to national attention by an environmental disaster in Times Beach, Missouri. The "beach" in the town's name derived from its idyllic location along the Meramec River—a feature that contributed to its downfall. In 1971, a local contractor sprayed a horse riding arena for dust control with waste oil contaminated with dioxin. Within a week, "bushel baskets" of birds were found dead and seventy-five horses died or had to be euthanized. The same contractor sprayed the town's unpaved roads with the contaminated oil to suppress dust. Nobody was paying attention.[29]

A little over a decade later, Times Beach landed on a list of one hundred sites possibly contaminated by dioxin. On December 3, 1982, EPA technicians collected soil samples from the town. Two days later, flooding forced most of the two thousand residents of Times Beach to evacuate. When the sampling results came back just before Christmas, dioxin levels were found to be one hundred times higher than the level regarded as safe for human exposure. Residents were in shock. Their homes had just been ravaged by flooding, and now they were told that their town was contaminated with a toxic chemical. The flooding may have washed some of the pollutant away, but also dispersed it into the flooded homes. The story was widely covered in the media, as day after day, technicians in "moon suits" showed up collecting samples. The government eventually purchased and subsequently demolished the entire town. In hindsight, many view the destruction of the town as an overreaction. Nevertheless, the contamination incident was blockbuster news at the time and illustrates the challenges in dealing in real-time with a controversial contaminant for which little is known about its effects.[30]

Dioxin was also a component of Agent Orange—a blend of "tactical herbicides" the U.S. military sprayed in Vietnam to destroy crops and remove dense tropical foliage that provided enemy cover. After years of controversy, the U.S. government now provides disability compensation for veterans with diseases associated with exposure to Agent Orange. There's no compensation for the Vietnamese who were exposed.

Chromium-6 is best known thanks to Julia Roberts's role as Erin Brockovich in her Oscar-winning portrayal of the David vs. Goliath victory over Pacific Gas & Electric. Chromium is used for a variety of purposes, including metal plating, paint, and corrosion inhibition in cooling towers. What complicates matters considerably is that there are two types of chromium—chromium-3 (which is good) and chromium-6 (which is bad). Chromium-6 is toxic and is a well-documented carcinogen, particularly when inhaled. Under the TSCA, restrictions on the use of chromium-6 center on commercial cooling systems and towers. Separately, debates continue over its regulation in drinking water.

"New" chemicals that came online after the TSCA was passed received more individual attention—but *more* is a relative term here, as opposed to basically none for all sixty-two thousand "existing chemicals." The way the law was framed hindered much in the way of regulation. First, the bar for establishing "unreasonable risk" was set very high. And second, the EPA had only ninety days to decide if it was going to investigate a certain chemical. The agency rarely had the toxicity data it needed because chemical companies weren't required to conduct toxicity testing unless a chemical was produced in very large quantities. This made for a classic chicken-and-egg predicament. The EPA needed evidence that a chemical posed a risk before it could require testing, and it wasn't going to get that evidence *without* testing. With its hands basically tied by the sheer number of chemicals that needed to be tested, the EPA relied on computer models and analogies to similar chemicals for assessing risk for the vast majority of these chemicals.

A major setback to the EPA's authority came in 1991, when a judge struck down the agency's gradual phase out of asbestos. The reasoning behind this decision was that the EPA had not considered all the alternative approaches that might be "less economically burdensome" to industry. This set up a nearly impossible task. The court upheld bans already in place for uses such as building materials, but overruled bans not yet in effect, such as brake drum linings. The EPA had invested millions of dollars and countless work hours into this effort. Nearly one hundred thousand pages of documents demonstrating the dangers had been compiled. Asbestos became the poster child of why congressional overhaul of the TSCA was urgently needed.[31]

By the 2000s, with no overhaul in sight, a patchwork of state regulations was of growing concern to chemical companies. Driven in large part by concerns about children, pregnant women, and other vulnerable populations, states were taking matters into their own hands and regulating chemicals with their own set of standards. Researchers were raising serious health questions about the safety of a range of chemicals, including flame retardants in furniture and bisphenols and phthalates in plastic water bottles and children's toys. Companies like Walmart were notifying suppliers that they would no longer purchase products that contained chemicals that were considered unsafe. Instead of "one size fits all," these efforts by states and retailers began to create a massive headache for chemical companies. The time was finally ripe for change when Obama took office and appointed Lisa Jackson, a chemical engineer, as EPA administrator.

After eight years of weak environmental enforcement under Bush, Jackson inherited a staff problem similar to what Ruckelshaus dealt with after Anne Gorsuch left. "Oftentimes we're in a meeting," Jackson explained, "and somebody starts telling me, 'Well, we already know what this official—usually a local official—really wants.' I tell them I don't want to know that. I want to know what the science says."[32] Jackson set a brisk pace away from the policies of the Bush era.

When Republicans took control of the House in 2010, the GOP chairman of the House Energy and Commerce Committee promised that he was going to bring Jackson in on such a frequent basis that she was going to need her own parking spot at the Capitol. Testifying dozens of times before hostile House committees, she was subjected to harsh questioning that verged on disrespectful. One coal industry official attacked her for waging a "regulatory jihad."[33]

Environmentalists praised her as fantastic and gutsy. She was given the *nom de guerre* of eco-warrior, which was literally true as her signing the endangerment finding helped start the Obama administration's so-called war on coal.[34] Under Jackson, the EPA pressed for limits on greenhouse gas emissions from vehicles and power plants, established the first standards for emission of mercury and other air toxics from power plants, finalized a rule reducing industrial pollution that crosses state borders, revoked a permit for the nation's largest mountaintop removal coal mine, and worked to overturn the infamous Bush "fill rule" that allowed mining companies to bury streams and lakes under mining rubble. This is only a partial list. Yet with all Jackson's work on climate and clean energy, the issue closest to her heart was reforming the TSCA.[35]

Lisa Jackson grew up in the Ninth Ward of New Orleans, near the toxic corridor known as "Cancer Alley." Thus, her commitment came from personal experience. With a master's degree in chemical engineering from Princeton, she attended college around the time the Love Canal debacle was unfolding. To Jackson's mind, this environmental catastrophe reinforced the need to protect vulnerable communities that lack political clout. There are a lot of them. In the United States, nearly three-fourths of Hispanics live in communities that fail to meet clean air standards, African Americans are more than twice as likely as whites to die from asthma, and Native Americans lack clean water at almost ten times the national rate.[36]

Jackson's motivation for reforming the TSCA came from getting out and talking to people. As she explained, "It came from trying to put my fingers on the pulse of what the average American cares about. I think there's huge grassroots concern and not just amongst environmentalists on this issue. I've talked to nurses, I've talked to religious leaders, I've talked to mother's groups, autism groups, you just name it, and everyone's worried about the same set of issues."[37] Jackson envisioned, and worked toward, a reformed TSCA that would require manufacturers to supply enough information to conclude that new and existing chemicals don't endanger public health or the environment.

It was not to be. Jackson resigned at the end of Obama's first term. Four years of malicious slander (and outright lies) against herself and the agency she served had certainly taken a toll, but nonetheless she had remained resolute to the principles of her job. Perhaps the straw that broke was a new tactic to get rid of her—the same one that was later perfected to bring down Hillary Clinton. Republicans in Congress had convinced the EPA's inspector general to open an investigation into Jackson's use of a secondary email account to conduct business inside the agency. Jackson explained she used the second account for everyday expediency because her public email address was widely known.[38]

Gina McCarthy, who succeeded Jackson as EPA administrator, continued working toward TSCA reform. During McCarthy's term, a chemical spill in West Virginia brought additional attention to the issue of chemical safety.

The first calls to the West Virginia Department of Environmental Protection started coming in around 8:15 in the morning. People were calling from Charleston (the state capital), reporting a licorice-like smell near the Elk River. *Licorice*? Yes, callers explained, a sort of sickening sweet smell.[39] It was January 9, 2014—the dead of winter in this mountain state. People were dashing from cars to buildings, not strolling along the river, so the smell must be strong. The Elk River is a tributary of the Kanawha River, known as Chemical Valley because of all the chemical processing and production plants lining the river valley. People here are used to all kinds of smells that would make many visitors gag, because the olfactory sense becomes dulled with constant exposure. State officials hit the road.

They went to where the complaints began and soon found the problem. About a mile above the main water intake for West Virginia American Water, the largest water utility in the state, were fourteen aging storage tanks owned by Freedom Industries, a producer of specialty chemicals. A large pool of clear liquid was soaking into the ground by the river, fed by a four-foot wide stream coming from a break in a concrete containment block. They traced the source back to two small holes in the floor of one of the tanks. There were no cleanup or serious containment measures underway. Instead, someone had stuck a cinder block and a fifty-pound bag of safety absorbent powder in front of the holes to try to stop the leak. "A Band-Aid approach," noted a state inspector. From the size of the roughly four-hundred-square-foot pool, it was clear this hadn't just happened.[40]

West Virginia American Water was aware of the chemical spill by noon but assumed that they could filter it. By four o'clock, their carbon filtration system could no longer handle the large amount of contamination in the water, and the chemical began getting into the water supply. By six in the evening, alerts went out to the utility's three hundred thousand customers (the entire Charleston metropolitan area and parts of nine counties) to stop using tap water for drinking, bathing, or cooking. Don't use it for anything, except flushing toilets.[41]

It was too late. The phone at West Virginia's poison control center was ringing nonstop with callers reporting a range of symptoms, including nausea and rashes. People were arriving at emergency rooms with symptoms of vomiting, trouble breathing, severe eye irritation, and skin blistering. A few had to be admitted.

Governor Earl Ray Tomblin declared a state of emergency and called in the West Virginia National Guard. President Obama declared a federal state of emergency. Tanker trucks with potable water began heading to affected areas. The Department of Homeland Security sent sixteen tractor trailers of bottled water to distribution centers in the Charleston metropolitan area.

For the next five days, the state's capital city almost came to a standstill. Businesses and schools were closed. Hospitals were taking emergency measures to conserve water. The university canceled classes. County offices were shuttered. More than three hundred people required medical treatment.[42] No one died, but at the same time, no one knew if there were long-term health effects. After five days, the ban began to be lifted, but the problem continued for weeks. Long after the city's drinking water had returned to "safe" levels, people could still smell and taste the toxic chemical in the water.

In December 2014, a federal grand jury indicted the four owners of Freedom Industries. According to the indictment, they spent money only to increase business and failed to make repairs, upgrades, and improvements necessary for environmental compliance and safety. They also had failed to develop a containment plan, should a tank leak.[43] In August 2015, the former president and co-owner of the company, Gary Southern, pled guilty to three charges. He faced up to three years in prison and a fine up to three hundred thousand dollars. Six months later, he was sentenced to one month in prison and a fine of twenty thousand dollars—a hand slap compared to his considerable wealth and the financial losses to the area.[44]

The Charleston spill had one notable feature that helped save the day. The contamination was discovered from the chemical odor—a leak of an odorless chemical would not have been detected so quickly. The chemical, 4-methylcyclohexane methanol (MCHM for short), was a relatively unknown chemical used for washing impurities off coal. Lack of knowledge about its properties severely hampered the emergency response. How concerned should people be? What are safe exposure levels? What are the long-term consequences of exposure? A second chemical, a mixture of polyglycol ethers, was also present but not reported by Freedom Industries until twelve days after the leak discovery.[45]

THE TSCA REFORMS

After years of debate and inaction, on June 22, 2016, President Obama signed the TSCA reforms—The Frank R. Lautenberg Chemical Safety for the 21st Century Act, to be precise—named for a former U.S. Senator from New Jersey who originally sponsored the act. Having gone through the Charleston spill (and contamination at Parkersburg—see next chapter), both Senators and all three Representatives from West Virginia voted in favor of the bill.

The revised TSCA strives to address the biggest problems with the original law, giving the EPA enhanced authority to require testing of both new and existing chemicals. Under the new strictures, a safety finding for a new chemical is required *before* it can enter the market. For chemicals already on the market, the EPA must establish a risk-based process to determine "high-priority" chemicals, and then complete a risk evaluation to determine their safety. The question of whether to regulate a chemical is evaluated against a purely health-based safety standard, thereby preventing a repeat of the cost to industry problems with asbestos. The EPA must consider vulnerable populations, such as children and pregnant women, in their assessment. Additional consumer information about chemicals is also made available by limiting a company's ability to claim information as confidential.

These are some of the major features of the reformed TSCA. After four decades, and at time of unprecedented political polarization, a Congressional majority agreed to a major overhaul of chemical regulations. Each side got something they wanted. The chemical industry got preemption from most new state regulations, and environmentalists got assurances that new chemicals would be evaluated on health and safety risks alone, not financial considerations. It all looked good, sounded good, and *was* good, but there was a catch— implementing the TSCA reforms was up to the Trump administration.

At the end of November 2016, a few weeks after Trump's election, nine senators wrote a letter to the presidential transition team urging the new administration to "maintain momentum" in implementing the TSCA reform. Among the nine were environmentally oriented Senators Tom Udall (D-NM) and Tom Carper (D-DE), but also three Republicans normally antagonistic to the EPA, including James Inhofe (R-OK), who had once compared the EPA to Nazi Germany's Gestapo.[46] It appeared that the reforms still had a bipartisan push behind them.

The goodwill continued during the first six months of the Trump administration. With the addition of more staff, the EPA was steadily reducing the backlog in new chemical reviews that had resulted from the law's new requirements. The review process was clearly more rigorous. Of 373 new chemicals reviewed, only eighty-one (22 percent) were given the green light to commercialize without certain restrictions—such as additional testing, bans on water discharge, or mandated use of personal protection equipment when handling the chemical. None were banned. According to Richard Denison, lead scientist at the Environmental Defense Fund, "If you look at those statistics—22 percent go onto the market [unrestricted]—that's 1 out of every 5, and it used to be the inverse of that. That's a pretty dramatic change."[47] Prior to this, the approval process for new chemicals was so lax that companies used to submit chemicals they didn't even plan to bring to market, Denison added.[48]

The honeymoon was short-lived. In August 2017, the EPA released its guidance document on how it would implement the TSCA reforms. The guidelines had been largely written by Nancy Beck who had come to the EPA from the American Chemical Council, the chemical industry's main trade association. With Pruitt's support, Beck took over writing the agency rules, pushing through industry-friendly changes.[49]

The Environmental Defense Fund, the primary environmental group promoting the TSCA reforms, protested that the EPA had watered down the intent of the reforms. A major concern was that the EPA was moving away from requirements that the agency consider all "intended, known or reasonably foreseen" uses. "Chemical manufacturers may produce a substance for a specific use," said Denison, "but once it's put on the market, it can end up being used in a wide variety of ways."[50] Under Beck's rules, uses beyond those identified by the manufacturer could be addressed later through a separate, less rigorous process. The fewer uses examined, the faster the review, and the greater the possibility that EPA staff would conclude the chemical is safe.

In addition to reviews of new chemicals, the EPA's other task was to replace the ad hoc (and ineffective) program of chemical reviews for existing chemicals with a mandated schedule to prioritize and evaluate existing chemicals. By now, the EPA's TSCA inventory listed over eighty-six thousand chemicals, although only about half of these were still on the market and about eighty-seven hundred of these had "meaningful production volumes," according to the American Chemistry Council.[51] Any way you look at it, this was still a large number.

Recognizing that it would take time to get started, Congress specifically called for faster action on three chemicals: TCE (a commonly used solvent) and two chemicals used in paint removers (NMP and methylene chloride). Among other health concerns, TCE is a known human carcinogen and NMP (widely used for graffiti removal) has been found to hinder fetal development and can cause miscarriage and stillbirth. Methylene chloride, however, stands out among the three.

Methylene chloride is widely used as an industrial solvent and as a paint stripper. Contained in numerous products sold in home improvement stores, the chemical has been linked to several kinds of cancer, as well as harming organs such as the lungs and kidneys. What drew particular attention, however, was its link to dozens of deaths nationwide, including twelve people between 2000 and 2011 who specialized in refinishing bathtubs.[52] Methylene chloride can quickly overwhelm workers and consumers, even when wearing masks or respirators, resulting in rapid asphyxiation and heart attacks. There is no way to significantly reduce exposure to methylene chloride when it is being used to strip paint, especially in a nonindustrial setting, such as homes. It takes a lot to do the job completely, and it is so volatile that it is difficult not to inhale some unless you're wearing a commercial full-face respirator—something do-it-yourselfers are unlikely to do. "Methylene chloride is arguably the most dangerous of all the solvents sold at Home Depot," said Dr. Josh Bloom, senior director of chemical and pharmaceutical sciences at the American Council on Science and Health (a conservative nonprofit that normally argues in favor of chemicals). "Chemists use it all of the time, but we do so in fume hoods."[53]

In January 2017, as one of the last actions taken by the EPA during the Obama administration, the agency proposed banning methylene chloride (and NMP) in paint strippers.[54] This was the first proposed ban by the EPA on uses of existing chemicals since its largely failed attempt to ban most uses of asbestos almost three decades earlier. "For the first time in a generation, we are able to restrict chemicals already in commerce that pose risks to public health and the environment," declared Jim Jones, assistant administrator of the EPA's Office of Chemical Safety and Pollution Prevention.[55] But the momentum was short-lived. In December 2017, Trump's EPA announced that the proposed ban was indefinitely postponed.[56]

The solvents industry agreed that methylene chloride is unsafe for bathtub and similar work but argued that the EPA action overstated the risks and would leave people without a good paint stripping alternative. The industry wanted a warning label instead of a ban. But how many people actually take the time to read warning labels, and then follow them? Nothing works better than methylene chloride at removing paint, but safer chemicals exist. And others can be developed. The European Union banned use of methylene chloride in paint strippers in 2010.

While the EPA was procrastinating, at least two more people died from exposure to methylene chloride. Drew Wynne, at thirty-one years old, was resurfacing the floor of a walk-in refrigerator at his cold brew coffee company in Charleston, South Carolina. Kevin Hartley, twenty-one years old, was a trained contractor who was refinishing a bathtub. Both men were wearing respirators. In May 2018, a meeting with family members was arranged after Rep. Frank Pallone (D-NJ), the top Democrat on the House Energy and Commerce Committee, asked Pruitt if he had anything to say to their families.

"I wanted to use Kevin's story to try to save more lives," said his mother, Wendy Hartley. Hartley, along with Cindy Wynne and her son Brian, met with Pruitt and several of his aides at EPA headquarters. The families brought photographs of the two men and explained what had happened. Pruitt "was very attentive to us," Cindy Wynne said immediately afterwards. "He was somewhat surprised when we showed him the cans from Lowe's," where her son had purchased the paint stripper. According to the families, Pruitt agreed that methylene chloride was a problem, but when pressed on whether he would finalize the ban, he did not make a commitment. Two days later, the EPA announced that it would finalize a methylene chloride rule but revealed few details and offered no timeline.[57]

Meanwhile, commercial outfits were under pressure from consumer advocacy groups to remove the products from their shelves. In May 2018, the same month that the family members were meeting with Pruitt, Lowe's announced that it would stop selling products containing methylene chloride—some nineteen products in all. Lowe's decision came after more than two hundred thousand consumers nationwide signed petitions urging the company to act. Sherwin-Williams, Home Depot, Walmart, and other retailers soon agreed to phase out the product.[58] The EPA continued to drag its feet until March 2019, when the agency finally announced that it would ban methylene chloride in consumer products (commercial operators could continue using the product as long as they underwent training).[59]

The concept behind the TSCA reforms was to move closer toward the precautionary principle in evaluating chemicals with a presumption of risk unless data prove otherwise. Methylene chloride was the first existing chemical to be evaluated under the TSCA reforms. The result is far from reassuring.[60]

10

THE FOREVER CHEMICALS

The reason that everybody likes planning is that nobody has to do anything.
—California Governor Jerry Brown

Chemistry has brought about amazing advances in consumer products. We cook in nonstick Teflon pans that all but clean themselves. We wear outdoor clothing that is waterproof, windproof, *and* breathable all at the same time. We buy carpets and furniture that are protected against stains. If our two-year-old spills her drink on the new couch, we just wipe it away—no fuss, no mess. It's "carefree living, built right in," boasts 3M, the manufacturer of Scotchgard. It all seems to be too good to be true. And it is.

The downside to these miracle products arises because they all have in common a class of chemicals known as PFAS (short for per- and polyfluoroalkyl substances). Just about everything that is nonstick, water-repellent, or stain-resistant—cookware, clothing, carpets and furniture, greaseproof fast-food wrappers, microwave popcorn bags, even foams for fighting oil-based fires—are all brought to us compliments of PFAS. Over the past two decades, PFAS have gone from virtually no recognition outside the chemical world to a household name for those who live near manufacturing facilities that use PFAS, on or next to military bases, or near civilian firefighter training sites.

The problems of PFAS arise from their chemical structure—a daisy chain of carbon atoms with carbon-fluorine (C-F) bonds, the strongest chemical bond in nature. The C-F bond is what makes PFAS so useful, but at the same time so long-lasting that they're known as "forever chemicals." Not a good thing for a bad chemical.

Of the more than four thousand PFAS, two have attracted the most attention—PFOA and PFOS (the "O" in the acronym helps distinguish these two chemicals from PFAS as a chemical group in the following discussion). PFOA was the first to gain national attention as a result of its use in the manufacture of Teflon, a product sometimes called "the most slippery substance on Earth."[1] DuPont, the manufacturer of Teflon, would prove just as slippery in defending PFOA.

PFAS are being found worldwide in the environment, wildlife, and people (almost every American has PFAS in their blood), but the public wake up to the dangers of PFOA came about because of the misfortunes of Wilbur Tennant, a cattle farmer in Parkersburg, West Virginia, where DuPont operated the world's first Teflon manufacturing plant. After the company bought land adjacent to Tennant's farm for a "non-hazardous" landfill, a creek that ran through the landfill into his pasture began to turn frothy and green colored. Tennant's cows became sickly. Within a few years, hundreds died, some succumbing to gruesome deaths.[2]

Tennant sought help locally but was spurned wherever he turned. Parkersburg was a company town—virtually everyone was connected one way or another to DuPont. Eventually, Robert Bilott, a corporate defense attorney in Cincinnati, took on the case as a favor to his grandmother who lived in the area. In preparing for the trial, Bilott came upon a letter sent to the U.S. Environmental Protection Agency (EPA) mentioning PFOA as a key chemical in the manufacture of Teflon. PFOA was not on any list of regulated substances, but Bilott's curiosity was aroused.

Through a court order, Bilott forced DuPont to share all documentation related to PFOA. Dozens of boxes containing thousands of unorganized documents arrived at his office, including private internal correspondence, medical and health reports, and confidential studies conducted by DuPont scientists. As he read through the documents, Bilott realized that he had a gold mine of incriminating material. "It was one of those things where you can't believe you're reading what you're reading," Bilott recalled. "It was the kind of stuff you always heard about happening, but you never thought you'd see written down."[3]

DuPont had conducted secret medical studies on the chemical for more than four decades. By the 1990s, DuPont knew that PFOA caused testicular, pancreatic, and liver cancer in lab animals, and possibly humans. In violation of the Toxic Substances Control Act, the company did not inform the EPA of these findings. Beginning in 1984, DuPont also had found the chemical in the tap water of nearby communities but failed to inform any water supplier for seventeen years.[4]

In 2001, Bilott initiated a class action suit on behalf of about seventy thousand people who had been drinking water from six public water districts and hundreds of private wells contaminated with PFOA. But he faced a legal dilemma—how could he argue that seventy thousand people had been poisoned by PFOA when the chemical wasn't on any federal or state list of contaminants? To get around this impediment, Bilott filed a medical monitoring claim that worked as follows: DuPont agreed to fund an independent panel of scientists to determine whether there was a "probable link" between PFOA and any diseases. If such links were found, DuPont would pay for medical monitoring requested by the exposed residents. Any of the class members who developed one of the linked diseases would be entitled to sue for personal injury. The catch was they couldn't do so until the study results were released.

As part of the settlement, DuPont paid seventy million dollars in damages upfront. Members of the class action suit voted to make each person's upfront cash award contingent on a medical examination and blood sample. Within months, nearly seventy thousand people were trading a sample of their blood for a check for four hundred dollars. DuPont also had agreed to fund the research without limitation. With comprehensive individual data on a large population and unlimited financial resources to carry out the study, the team of scientists had hit an epidemiological jackpot.

In 2011, after seven years and thirty-five million dollars spent, the scientific panel began to release its findings. They concluded that there was a "probable link" between PFOA and six health problems: kidney cancer, testicular cancer, thyroid disease, high cholesterol, pregnancy-induced hypertension, and ulcerative colitis. After the science panel released its findings, plaintiffs with these health problems filed personal injury lawsuits against DuPont. After a few bellwether cases resulted in large settlements, DuPont and its spin-off company, Chemours, settled with 3,550 plaintiffs for $671 million.[5] Subsequent studies by others have since suggested links to other health problems, including obesity, decreased fertility, developmental delays in fetuses and children, immune suppression, endocrine disruption, and reduced effectiveness of children's vaccines—but these were not part of the settlement agreement.

In 2005, DuPont reached a $16.5 million settlement with the EPA for its decades-long cover up of PFOA. This was the largest civil administrative penalty the EPA had obtained in its history but represented less than 2 percent of Dupont's profits from using PFOA for just that year.[6] In 2006, the EPA helped broker a deal with DuPont and seven other companies to phase out production of PFOA by 2015. But the environmental and health impacts of PFOA and other forever chemicals are far from over.

Parkersburg, West Virginia, is not alone. Other towns near companies using PFAS have been forced to shut down their water supplies or install expensive water treatment. Among these are Decatur, Alabama; Hoosick Falls and Petersburgh, New York; North Bennington, Vermont; Merrimack, New Hampshire; and Parchment, Michigan. Wolverine, the maker of popular footwear brands like Hush Puppies, used 3M's Scotchgard in waterproofing its shoes. The wastes were dumped at numerous, and largely undocumented, sites around the company's home town of Rockford, Michigan. When PFAS began showing up in private wells, no one knew where these chemicals would show up next.[7] In another case, 3M settled with the state of Minnesota for $850 million to address PFAS contamination—the largest environmental lawsuit in the state's history.[8]

The most prevalent contamination problems have come from firefighting foams that have been used at military and municipal fire training areas for decades. These foams work spectacularly well at quenching an oil fire by blanketing the fuel, cooling the surface, and suppressing release of flammable vapors to prevent re-ignition. The quick-acting firefighting foams save lives, including those of firefighters. However, for years, the standard practice was to drain the PFAS waste from firefighting exercises into unlined pits. Consequently, groundwater near virtually every military air base in the country is contaminated by PFAS.

By 2017, the Department of Defense had identified 401 military installations with known (or suspected) releases of PFOS or PFOA. Studies by the Pentagon found PFAS contamination in drinking water or groundwater in at least 126 locations, including systems that supply water to tens of thousands of people on the bases and in nearby neighborhoods. The military has been proactive in providing temporary alternative drinking water supplies and installing treatment for many of these contaminated systems, along with many private wells. However, with billions of dollars at stake—and reminiscent of the debates over perchlorate—the Pentagon has pressured the EPA to weaken standards for groundwater cleanup.[9]

Addressing the overall PFAS problem has been fraught with difficulties. It's in groundwater, surface water, drinking water, food, consumer products, effluent from wastewater treatment plants, landfills, biosolids used for fertilizer and soil amendment, fish, and wildlife. Despite studying PFOA and PFOS for two decades, toxicologists are still struggling to work out how PFAS cause problems in the body.[10] There's also an alphabet soup of PFAS being used worldwide. Many PFAS "precursors" may not be highly toxic in themselves, but they naturally degrade to PFOA, PFOS, and other toxic PFAS.

Developing guidelines for safe levels of PFAS in drinking water has been painstakingly slow. Bilott informed the EPA about his findings in March 2001. In 2002, the EPA initiated a priority review under the Toxic Substances Control Act. But after 2006, with agreements in place to phase out PFOA and PFOS, the EPA largely dropped the ball by failing to allocate sufficient funding for research—further delaying an already slow regulatory process. [11]

In 2009, the EPA set provisional (short-term exposure) health advisories for PFOA and PFOS at four hundred and two hundred parts per *trillion*, respectively. In 2016, the EPA issued a formal (long-term exposure) health advisory of seventy parts per trillion for PFOA and PFOS combined, much lower than the earlier provisional health advisories. But these levels are still just advisory, not enforceable. A health advisory of seventy parts per trillion equals 0.07 parts per billion, much lower than the few parts per billion standards for trichloroethylene, tetrachloroethylene, and other organic compounds. Some researchers have even suggested that the safe level may be less than one part per trillion. [12] The delays in developing a federal drinking water standard have caused many states to adopt their own advisories and standards.

It wasn't until 2013 to 2015 that the first nationwide evaluation of PFOA, PFOS, and four other PFAS in drinking water took place under the Unregulated Contaminant Monitoring Rule of the Safe Drinking Water Act. Based on those data, some six million Americans (about 2 percent of the population) obtain their drinking water from sources that exceed the EPA health guidelines for PFOA and PFOS. [13] This number does not include people with contaminated private wells.

In 2018, the uncertainty about the safe level of PFAS took another turn when the Agency for Toxic Substances and Disease Registry (ATSDR; part of the Department of Health and Human Services) found that exposure to PFOA and PFOS in drinking water could be harmful at levels seven to ten times lower than what the EPA had estimated for the two *combined*. Toxicological profiles are routinely issued by the ATSDR to inform agencies like the EPA as they consider regulations. But what made these profiles particularly newsworthy was that they came via a Freedom of Information Act request by the Union of Concerned Scientists. The response to the

Freedom of Information Act request revealed email correspondence among the White House, EPA, and Department of Defense suggesting collusion in holding back release of the ATSDR findings. An unnamed White House aide warned that this could be a "public relations nightmare" for the Trump administration.[14]

With PFOA and PFOS being phased out, companies have switched to PFAS with shorter carbon chains. These compounds exit the body quicker and appear to be less toxic. But they are also more mobile in groundwater and more difficult to remove in drinking water treatment. Currently, there is relatively little data on the environmental persistence and toxicity of these shorter-chain compounds.[15] Accepting these product substitutions for widespread use without a more thorough review by independent experts has been widely criticized as failing to learn the lessons from PFOA and PFOS.

In the absence of regulations, health advisories provide interim guidance on safe levels of a contaminant. But they also can lead to confusion and present major communication challenges. It's hard for people to understand that a health advisory doesn't carry the same weight as a drinking water standard. A health advisory is simply "informal technical guidance," which a utility can choose to use or not. So, people might ask if the EPA has issued a health advisory for a particular contaminant, then why isn't the water utility doing more to test and treat for it? And if there's need for a health advisory, then why doesn't the EPA regulate it? These are all perfectly logical questions. There's another problem: With less rigorous science behind them, health advisories are more subject to change, leaving utilities and local officials holding the bag to explain why yesterday's safe level is no longer considered safe. In short, health advisories are a classic catch-22—damned if you do and damned if you don't. This is a familiar predicament if you work for the EPA.

Hoosick Falls, New York, is an example of the confusion that can result when a drinking water supply becomes contaminated by an unregulated chemical, compounded by evolving information on safe limits. The Village of Hoosick Falls is located in the Town of Hoosick, a rural community thirty miles northeast of Albany, New York, almost on the Vermont state line. The town has a population of sixty-seven hundred with about half living in the village. With its tree-lined streets and Victorian houses, Hoosick Falls is like stepping back in time. The village is best known as the resting place of Grandma Moses, who began her career painting bucolic rural scenes at age seventy-eight.

The village water supply is served by three wells. In August 2014, Michael Hickey, a former village trustee, requested that water samples from the municipal water system be analyzed for the presence of PFOA. Near the village are two Saint-Gobain plants, a French company manufacturing Teflon-coated material, as well as a shuttered Honeywell plant that had used PFOA for decades. Hickey knew several people who had died from rare types of cancer, including his father, who had worked at the Saint-Gobain plant. Michael Hickey was searching for answers.[16] The New York State Department of Health, which oversees operations at the village water treatment plant, responded that the water was safe, and it wasn't necessary to test for PFOA. The Village Board decided to obtain the water samples anyway.

The sampling results showed PFOA levels ranging from 180 to 540 parts per trillion. At the time, the EPA's provisional drinking water health advisory was four hundred parts per trillion. Although some of the samples were above this level, PFOA was an unregulated compound, so no action was triggered. Hoosick Falls is located more than ten miles from the closest alternative treated water source and did not have a contingency water supply available.

In December 2014, the villagers' water bills included a letter from the mayor about the test results. They were assured that the water supply continued "to meet and exceed all County, State and Federal standards for public health safety. If that was not the case, the Rensselaer County Public Health Department would intervene immediately." Despite these assurances, the next three years became a traumatic rollercoaster ride for the village and nearby residents. Imagine finding yourself on the receiving end of the following chronology.[17]

On April 1, 2015 (121 days after the mayor's first letter), a second letter from the mayor was included with the water bill. This one began "Hello Folks" and contained the same nothing-to-be-worried-about message, along with updates on actions so far. A pilot study showed that granulated activated carbon (GAC) was an effective treatment for PFOA in the municipal water supply, and the village began to seek funding from the state for the treatment system.

In August 2015 (Day 243+), test results from sampling nearby private wells revealed that PFOA was found in three of the eleven wells tested. As a result, the state planned to sample additional private wells. A third letter from the mayor included with water bills focused on efforts to obtain funding for the GAC treatment system, which was estimated to cost more than two million dollars to install.

In September 2015 (Day 274+), the state Environmental Facilities Corporation notified the village that it was ineligible for funding for the GAC system because funding was targeted to communities contaminated with regulated chemicals. Unsuccessful at obtaining funding for the treatment system, the mayor pushed for a meeting with Governor Andrew Cuomo.

In November 2015, Saint-Gobain agreed to fund the GAC treatment system and, until operational, provide free bottled water for residents. The bottled water program began at the end of the month, almost one year to the day after the mayor's first letter. Meanwhile, the EPA sent a letter to the village mayor recommending that municipal water not be used for drinking or cooking.

On December 1, 2015 (Day 365), the mayor's fourth letter was sent out with water bills, announcing the free bottled water program. The letter didn't mention the EPA's recommendation not to use the water for drinking or cooking. On December 17, the EPA stated emphatically its recommendation that "people NOT drink the water from the Hoosick Falls public water supply or use it for cooking." As a precautionary measure, the EPA also recommended that children or people with skin conditions avoid prolonged contact with

PFOA-contaminated water, such as long showers or baths. The next day, a fifth letter was mailed to all municipal water customers with the EPA's recommendations.

On January 27, 2016 (422 days), Governor Cuomo issued an emergency regulation to classify PFOA as a hazardous substance. The governor committed to allocating significant state resources to investigate the source of PFOA, to conduct a Health Risk Analysis for establishing a state drinking water guidance level for PFOA, to retest private wells, and to immediately install filtration systems at the school and other community gathering places. A state hotline was also established.[18] On January 28, the EPA recommended that any resident with a private well at which PFOA had been detected at levels greater than one hundred parts per trillion (one-quarter the earlier provisional advisory level) should not use the water for drinking or cooking. EPA Spokeswoman Mary Mears said this recommendation was made "out of an abundance of caution."[19] The EPA also recommended that any resident with a private well that had not yet been tested ask the state Department of Health to do so. In the meantime, they should use the free bottled water.

In February 2016, the latest sampling results indicated that forty-two of the 145 private and municipal wells tested (nearly a third) had PFOA levels above the EPA's most recent cutoff level of one hundred parts per trillion. The state identified Saint-Gobain and Honeywell as the potentially responsible parties for the contamination and notified them that they would be held accountable for remediation. Students at the Hoosick Falls Central School District held a press conference to voice their concerns and urge New York State to find an alternative water supply. Later that day, Governor Cuomo announced that the state had begun planning for a possible alternate water supply for the village. The governor also announced that the state would allocate ten million dollars to purchase and install water filtration systems for approximately fifteen hundred private wells. Free blood tests would also be made available for residents. The state Department of Financial Services set up a command center to help residents with mortgage issues after two local financial institutions announced they would no longer issue mortgages in light of the PFOA issue. On February 25 (Day 451), Cuomo tried to rewrite history by claiming: "We've been very active in Hoosick Falls from Day 1."[20]

On March 13 (468 days), Governor Cuomo (who prides himself on fast and aggressive responses to storms and other disasters) made his first visit to Hoosick Falls. He announced that the new filtration system was successfully removing PFOA to non-detectable levels (less than two parts per trillion). However, residents were still cautioned not to use tap water for drinking or cooking until a full flush of the local water system had been completed. Cuomo promised residents that there would be continued long-term action on the water problem. He also expressed his frustration with the EPA. "We think the EPA should set a number, and whatever that number is we'll follow," the governor said. "But we need the number."[21] On March 30 (485 days), the no-drink advisory was finally lifted and village residents could "use the water for any and all uses, including drinking or cooking." The two local financial institutions reinstated mortgage programs.

On May 19, 2016 (Day 535), the EPA announced a new lifetime health advisory of seventy parts per trillion (for PFOA and PFOS combined). According to the EPA, the new advisory level "offers a margin of protection for all Americans throughout their life from adverse health effects resulting from exposure to PFOA and PFOS in drinking water." The advisory is non-enforceable. The large change in the advisory level added to the confusion at Hoosick Falls and elsewhere.

On September 7, 2016 (Day 646), the EPA proposed adding the Saint-Gobain facility to the Superfund National Priorities List. The state initiated a study to collect and analyze fish from local water bodies potentially contaminated with PFOA.

In June 2017, Saint-Gobain officials announced that a groundwater sample beneath their site showed PFOA levels at 130,000 parts per trillion. "These may be the highest levels of groundwater contamination for PFOA identified in the nation to date," commented the former EPA regional director.[22]

On July 31, 2017 (Day 973), the EPA announced addition of the Saint-Gobain site to the Superfund National Priorities List.[23] The state announced that fish in Thayer Pond, one of the waterbodies tested, have elevated levels of PFOA. A catch and release advisory was issued for the pond.[24]

On September 1, 2017 (Day 1,005), the free bottled water program ended as the GAC filters continued to result in levels of PFOA in the finished water that were below the detection limit. Understandably, residents continued to be wary of their drinking water and to demand a clean alternate source.

A few months earlier, eighth-grader Harmony Bishop had written an op-ed in the local newspaper expressing the frustration felt by many. The letter began "Thanks to my mom and a bunch of other moms in Hoosick Falls, Gov. Andrew Cuomo and state legislators created a water quality council." She noted that the council had not yet met, despite promises to do so. "I'm just curious," Harmony asked, "Is knowing something is due but promising to hand it in late or never handing in at all allowed at my school, too? Or is this something only politicians get away with?" Then she continued, "At school now, water is a pretty common topic. It's funny; I bet most teenagers don't talk to friends about water quality. In Hoosick Falls, we do. Just imagine being scared of the water coming out of your faucet. Or worrying that someday, your mom or dad or brother or sister may get cancer."[25]

From the local to the federal level, the PFOA contamination of Hoosick Falls' water supply was not the government's finest hour in responding to the crisis. The public lost confidence in officials, who started off by denying the problem and dragged their feet in facing up to the dangers. The GAC treatment system solved the town's water supply problem with PFOA and has the advantage of removing other contaminants, yet residents understandably see this as a short-term fix. There's also the problem of the many private wells that require installation and continued maintenance of treatment systems for the indefinite future.

The overall PFAS story continues to develop across the country and will do so for a long time to come. In February 2019, under increasing pressure from Congress, the EPA released a long-awaited PFAS Action Plan.[26] The agency vowed by year's end to begin the lengthy process of setting enforceable drinking water limits for PFOA and PFOS. The EPA also promised to issue interim groundwater cleanup recommendations, require more testing for PFAS in public water systems, undertake more research on the health effects of less studied compounds, and better communicate the risks to communities around the country. The proof will be if these promises are kept. "It's just disheartening," said Michael Hickey, whose persistence led to the discovery of PFOA contamination at Hoosick Falls. "These are real illnesses that happen every day. This EPA doesn't understand the severity and what it actually does to a small town when there's this kind of contamination."[27]

11

SUPERFUND

I thought he was saying we needed to get off business's back. I thought he meant all those forms people have to fill out. I never thought of it as applying to things like protecting public health and the environment. I've voted for every Republican presidential candidate since 1960. It's going to be a different story this time.
—Clyde Wallace, Cheraw, South Carolina, who voted for Reagan, speaking in 1983 while battling a hazardous waste landfill[1]

Imagine living in a neighborhood where you increasingly hear talk about serious health problems, and there's a startling similarity— family members with cancer and neurological disorders, children having recurrent rashes all over their bodies. And the number is growing. People begin to think about that long-shuttered factory by the river. Former employees recall the sloppy practices at the plant. There are other suspects—a nearby dry cleaner and some small businesses that handle toxic chemicals. The neighborhood gets more and more alarmed, and angry. Who can they turn to for help? The factory long ago went bankrupt and the owners are long gone. Local officials have no idea how to deal with the issue and are just trying to calm everyone down. Then word gets out, and no one can sell their home. More likely than not, this is a poorer neighborhood where people have few options. They've always lived and worked here. Where would they go, even if they could?

A decade after the U.S. Environmental Protection Agency (EPA) was created, dealing with such problems was still a major gap in environmental protection. Then came Love Canal. In 1978, the discovery of hazardous chemicals buried beneath houses and an elementary school in the Love Canal neighborhood of Niagara Falls, New York, made toxic contamination frontpage news nationwide. This was just the beginning, as contamination began to be found at sites across the country. Worse yet, toxic chemicals that had been indiscriminately dumped for decades were turning up in drinking water wells. The public's initial indignation turned to fear at the prospect of their drinking water being laced with odorless and tasteless toxic chemicals from these "ticking time bombs."

In 1980, Congress passed the Comprehensive Environmental Response, Compensation, and Liability Act—otherwise known as Superfund because of the price tag. Superfund's primary goals are to clean up contaminated sites and make those responsible for the contamination pay for the clean-up costs. If the "potentially responsible parties" cannot be identified, no longer exist, or cannot afford cleanup costs, then the EPA finances the cleanup from a trust fund and later recovers those funds from the potentially responsible parties. In other words, cleanup does not have to wait for long, drawnout lawsuits to be resolved. The trust fund can be used to address both emergency removal of hazardous substances and longer-term cleanups. The state where the site is located provides some cost-sharing with the federal trust fund.

Despite its name, Superfund has been chronically underfunded for decades. Under the principle that "the polluter pays," the trust fund initially was paid for by taxes on oil and chemical companies. These taxes expired in 1995, when Newt Gingrich and the Republicans took charge of the House. They have never been reinstated, leaving American taxpayers to foot the bill through annual appropriations to the EPA. Further hampering the Superfund program, these appropriations were cut by nearly half between 1999 and 2013 and continue to decline.[2] The reduced funding affects site cleanup, as well as the EPA's ability to make polluters pay. For every dollar the EPA has expended on enforcement, eight dollars have been collected from responsible private parties toward cleanup work.[3]

Industry and businesses are not the only culprits. By 2012, more than thirty billion dollars had been spent cleaning up hazardous chemicals at military bases.[4] Furthermore, the Energy Department's cleanup of contamination from nuclear weapons production is the largest environmental clean-up program in the world.[5] These federal agencies rely on their own (taxpayer-funded) appropriations for cleanup. The EPA and states provide oversight and enforcement.

Superfund is one of the EPA's most controversial environmental programs, criticized for lengthy delays, high costs, and limited accomplishments.[6] The program also got off to a slow and rocky start. Under Reagan, the first president to oversee Superfund, only a few sites were cleaned up and minimal funds were recovered from responsible polluters. Rita Lavelle, the first head of Superfund, was sentenced to six months in prison and fined ten thousand dollars for perjury and obstructing a Congressional investigation of her former employer, Aerojet.[7] Nonetheless, Superfund "would have presented severe challenges even to the most able, law abiding, and aggressive administrator," according to political scientist Marc Landy.[8] Frustrated with the pace of cleanup and the Reagan administration's lax implementation, Congress passed amendments to strengthen the act in 1986. At the same time, taxes to support the trust fund were increased.

Superfund's liability structure is intended to capture all parties that *may* have had *some* involvement in contamination of a site. In other words, the EPA can cast a wide net to assign liability. The idea is to minimize the costs to the general taxpayer. Past owners and operators of contaminated sites can be held financially responsible even if the wastes were disposed of legally at the time. Current owners can be held financially responsible even if they unwittingly inherited the problem. Potentially responsible parties not only include large industrial firms, but municipalities, hospitals, and small businesses. Consistent with litigation the world over, the EPA focuses on those with deep pockets for cost recovery. As a result, controversies about who picks up the tab for someone else's pollution is a constant battle—and a lucrative source of income for lawyers.

To become eligible for long-term Superfund funding, a site has to make it onto the EPA National Priorities List. Candidates are scored by a hazard ranking system based on the risks they pose to human health and the environment. Those that score high enough get on the list. There are currently more than thirteen hundred Superfund sites nationwide. All told, about one out of every six Americans lives within three miles of a Superfund site.[9] To illustrate some of the challenges, let's take a look at two areas—the Superfund complex associated with Anaconda's mining and smelters in Montana, and groundwater contamination in southern California.

The legacy of Anaconda's smelters in Montana, where Teddy Roosevelt lost his battle to curb the company's toxic air emissions in the early 1900s (see chapter 9), is a massive undertaking to clean up a century of widespread pollution. Until the 1960s, Anaconda virtually owned the state of Montana, including five of the state's six newspapers. A tell-tale anecdote harkens back to the 1920s, when an Anaconda lobbyist reportedly would throw rolls of cash into the hotel rooms of sleeping legislators with a note attached telling them which way to vote on bills the following day.[10] A rapid decline in the company's fortunes began in 1971, when Chile's socialist government nationalized Anaconda's operations—the primary source of the company's revenue. Falling copper prices worldwide compounded Anaconda's woes. In 1977, Atlantic Richfield Co. (ARCO) bought the struggling company. Three years later, ARCO abruptly closed the last Anaconda smelter, throwing more than a thousand people out of work.[11] The Butte open-pit copper mine closed two years later.

In a case of bad timing, ARCO had purchased Anaconda three years before Superfund was enacted. When the EPA added Anaconda's copper mining and smelter legacies to the Superfund National

Priorities List in 1983, ARCO found itself on the hook for the nation's largest Superfund complex. Over thirty-five years and more than a billion dollars later, ARCO is still cleaning up the mess.

A century of air emissions and shoddy waste disposal practices by the Anaconda smelters left a legacy of more than three hundred square miles of soil and water contaminated with heavy metals. Cleaning up the immediate areas around the smelters became the first order of business, eventually resulting in a world-class golf course designed by Jack Nicklaus. There was also a successful "Save the Stack" initiative by local citizens, whereby the 585-foot-tall smoke stack (of the most recent smelter) was preserved as a state park. This iconic landmark could almost contain the Washington Monument. However, unlike the Washington Monument, the stack has to be viewed from about a mile away because the surrounding arsenic-laden soils are still hazardous.

Cleanup of residential and commercial properties in the surrounding area has been slower and more problematic. Beginning in 2003, ARCO sampled more than seventeen hundred residential yards. Over 350 had been contaminated by arsenic and required cleanup. This cleanup effort failed to include lead, another common contaminant from copper smelters. As a result, ARCO had to return to remediate about one thousand homes, as well as school grounds and interiors. There were yet other surprises. In 2017, tests in a city park found elevated levels of arsenic and lead in the children's sandbox.[12]

Butte, once the home of the "Richest Hill on Earth," is twenty-five miles southeast of the Anaconda smelter and is part of this huge Superfund complex. The town's most notorious feature, Berkeley Pit, is a hole in the ground about seventeen hundred feet deep and a mile and a half wide. It remains as the legacy of one of the world's largest open-pit copper mines. When ARCO shut down the mining operation, they also shut off the pumps that kept out water. The pit began filling with highly acidic water laced with toxic metals and is now the deepest lake in Montana. By around 2022, the water is predicted to exceed a critical level where it could flow through shallow alluvium, contaminating groundwater and Silver Bow

Creek, a tributary of the Clark Fork River. To prevent this from happening, a treatment plant has been built to pump and treat water from the pit to be discharged into Silver Bow Creek or used elsewhere. The operation will need to continue *in perpetuity*.

Berkeley Pit is both a Superfund site and a tourist destination. For two dollars, visitors can have a bird's eye view of the toxic lake. The lake is also a death trap for birds. In 1995, a flock of snow geese landed on the lake. Bad weather and dense fog prevented them from leaving for several days. When the weather cleared, 342 snow geese were found dead. To keep this from happening again, a "waterfowl hazing program" uses boats, spotlights, shotgun and rifle noises, fireworks, and noise-emitting electronic devices to scare away the birds. However, this wasn't enough of a deterrent in 2016, when a major storm pushed huge flocks of late migrating snow geese into the area. Employees managed to scare thousands away, but an estimated three thousand to four thousand birds died from ingesting lake water. Horrified witnesses described the lake as "seven hundred acres of white birds."[13] Newer efforts are experimenting with drones and a handheld laser that can shine a green beam all the way to the other side of the pit.[14]

A significant Superfund accomplishment in Butte dealt with lead poisoning. After a number of studies found that almost 10 percent of Butte's children had lead levels in their blood that exceeded the safety guideline, lead was removed from attics, yards, and paint from more than one thousand homes. After nearly twenty-five years of work, it is now rare for a child in Butte to test higher than the lead safety guideline.[15]

Another accomplishment is the cleanup and restoration of a twenty-six-mile stretch of Silver Bow Creek. This effort is considered the largest project of its kind ever undertaken in the United States. For almost a century, tailings and other mine wastes were dumped indiscriminately along the creek's floodplain, creating a large area completely devoid of vegetation and wildlife. Periodic floods carried the toxic-laden sediments into the creek, which ran red and lifeless.

In 1999, ARCO finally settled with the state and EPA for cleaning up Silver Bow Creek. Additional funds were available from a separate suit by Montana against ARCO for natural resources damages of the entire Upper Clark Fork River Basin.[16] Over the next sixteen years, the state of Montana undertook the daunting task to temporarily relocate Silver Bow Creek, remove the tailings from the stream bed, and then put the stream back. Meanders, wetlands, and other features were added in order to restore the stream's ecological health. "We rebuilt the entire stream and floodplain," noted Joel Chavez, Montana's project manager. More than five million cubic yards of tailings and soils laden with heavy metals were removed from the creek and floodplain. Over a million willow trees and close to two million wetland herbs were planted. A greenway for walking and bicycling was constructed along its entire length. And most importantly to Montanans, after a gap of one hundred years, the fish are back.[17]

The Superfund complex continues 120 miles downstream on the Clark Fork River, all the way to Milltown Dam. Just months after the dam was completed in 1908, the largest flood on record for the Clark Fork washed untold tons of mine wastes into the reservoir. The dam was added to the Superfund complex after reservoir sediments were discovered to be leaching arsenic into the local drinking water aquifer.

Whether to remove the dam was controversial until the winter of 1996, when a fourteen-foot-thick ice jam threatened to take out part of the dam. The emergency release of reservoir water to make room for the ice scoured contaminated bottom sediment, sending it over the dam and killing fish downstream. The next spring, biologists reported nearly a two-thirds reduction in catchable rainbow trout.[18] This event also created concerns about dam safety for the residents of Missoula not far downstream. For a price tag of $120 million, the contaminated sediments and the dam are now gone.

Despite more than three decades as a Superfund site, clean-up work will continue for years—and essentially forever at Berkeley Pit. For some streams, meeting state aquatic life standards has been declared technically impracticable because of the widespread contamination. A new chapter in the cleanup is planned to begin in 2019, addressing contamination from about five hundred mine dumps west and north of Butte.[19] One challenge is determining the responsible party for each of these, which may not be ARCO. Cleanup of mining-impacted areas in the Butte urban river corridor also remains controversial. Activists groups have played a key role in keeping the pressure on all parties in these various efforts. Finally, there's the town of Opportunity, established by Anaconda in 1914 as a model community to demonstrate that people could raise crops and livestock in this heavily polluted area. Opportunity has failed to live up to its name, as the small town became a dumping ground for the region's excavated mine wastes.[20]

The Anaconda smelter, Silver Bow Creek/Butte, and Milltown Dam/Clark Fork River sites are often referred to as the nation's largest Superfund complex. There is, however, a competitor for a similar dubious distinction. The San Gabriel Valley in eastern Los Angeles County claims to be the largest Superfund site. The matter is settled by a nuance of phraseology. Montana has the largest Superfund "complex," whereas the San Gabriel Valley is the largest Superfund "site."

Following the outbreak of World War II, southern California experienced rapid growth, largely because of massive funding for military bases, bomb-making plants, and aerospace and electronics industries. This industrial growth coincided with the growing use of chlorinated solvents and other synthetic organic chemicals for cleaning machinery and other uses. After decades of poor handling and disposal practices, the San Gabriel Valley is besieged with widespread groundwater contamination. Groundwater is a critical resource in this region, particularly as imported water from northern California and the Colorado River becomes more expensive and less reliable in the face of climate change and increasing water demands.

The aquifer underlying the San Gabriel Valley is the primary source of drinking water for more than a million people. When the area was added to the Superfund National Priorities List in 1984, scores of wells were found to be contaminated with high concentrations of volatile organic compounds, including perchloroethylene (PCE) and trichloroethylene (TCE). Perchlorate and NDMA (a super toxic, mobile, and difficult to treat chemical) were later found. In contrast to ARCO being the responsible party for most of the Montana Anaconda Superfund complex, more than one hundred potentially responsible parties have contributed to soil and groundwater contamination in the San Gabriel Valley.

In 1993, the San Gabriel Basin Water Quality Authority was created to regulate pumping and work with the EPA to seek funding for a solution. Advanced groundwater treatment systems now supply clean drinking water to valley residents. The ongoing multi-decade effort required an innovative approach that included the first-ever perchlorate treatment facility for drinking water supply. As of May 2017, more than forty-five tons of contaminants had been removed from the groundwater and another forty tons of contaminants from the soil at industrial facilities with most costs borne by the polluting industries.[21] Treatment will likely continue for at least another fifty to sixty years. "Until it is done," says Kenneth Manning, executive director of the San Gabriel Basin Water Quality Authority.[22]

The problem is not restricted to the San Gabriel Valley. Groundwater in the neighboring San Fernando Valley once provided drinking water to more than eight hundred thousand residents of Los Angeles, Burbank, and nearby cities. In the 1980s, more than half the water supply wells were shut down because of groundwater contamination by chlorinated solvents, primarily PCE and TCE. Chromium-6 and other contaminants were later added to the list. After decades of limited progress in groundwater cleanup, the Los Angeles Department of Water and Power plans to build the world's largest groundwater treatment center in the San Fernando Valley.[23]

As water resources become increasingly stressed nationwide, the San Gabriel and San Fernando Valleys are a warning sign about the consequences of groundwater contamination by toxic chemicals. Groundwater is a precious local resource in both areas, but it can be used only with extremely costly water treatment that has no end in sight.

Superfund and the litigation and expenses of cleaning up contamination resulted in unanticipated consequences. Fearing future liabilities for contaminant cleanup, lenders were becoming increasingly hesitant to provide loans to redevelop blighted and abandoned industrial properties. There are also many more contaminated industrial areas than can possibly be remediated under Superfund. To address these issues, the EPA launched a "Brownfields" initiative in the 1990s to help local governments revitalize abandoned or underutilized industrial sites. These programs encourage private parties to remediate such sites voluntarily by providing liability protection for good faith efforts. Ideally, they also help disadvantaged communities and foster job creation.

Emeryville, located between Oakland and Berkeley, California, was an early Brownfields pilot project. During the 1970s, industry largely abandoned the city. By the mid-1990s, more than 230 acres within Emeryville were underused or vacant, and more than 90 percent of this land had significant soil or groundwater contamination, or both. The EPA awarded Emeryville a two hundred thousand dollar grant in 1996 and worked with the city to target ten brownfields areas deemed suitable for redevelopment. On one of these sites, a private corporation purchased the property and constructed two hundred units of mixed-income housing. Another brownfield site was purchased by one of the country's largest biotechnology firms for its new headquarters. As of 2006, Emeryville had leveraged more than $640 million in clean-up and redevelopment funding from the private sector through its Brownfields program.[24]

Such early successes led to widespread calls for congressional legislation to codify the Brownfields program into law. In 2002, President George W. Bush signed the Small Business Liability Relief and Brownfields Revitalization Act with bipartisan support. The Brownfields Law provides important protections from Superfund liability to landowners who meet certain criteria. States followed suit by adopting laws that eased the threat of liability for voluntary clean-up programs. EPA grants through the Brownfields program have leveraged about seventeen dollars for each EPA dollar. As of

February 2019, an estimated 145,000 jobs have been created and more than seven thousand properties made ready for reuse.[25]

Revitalizing inner city brownfields is considered one of the EPA's major accomplishments, but the picture is not completely rosy. By letting insurers too easily off the hook, considerable contamination may be left behind. In addition, community stakeholders have repeatedly voiced concerns that the program may unintentionally exacerbate gentrification and displacement of low-income and minority communities. In 2006, the EPA's federal advisory committee on environmental justice published a widely distributed report highlighting this issue.[26] The report led the EPA to be more conscientious of potential unintended impacts throughout the Brownfields revitalization process.[27]

These examples illustrate many of the challenges that the EPA faces, including limited funding and staff, how to address myriad public concerns in an intense and emotional environment, foot-dragging by companies deemed responsible for contamination, and a basic lack of data about some sites and the health hazards they pose. Meanwhile, those affected by contamination—as well as those undertaking cleanup—want certainty and timely decisions.

Despite shrinking budgets, Superfund has been relatively successful in reducing human exposure to toxic contaminants. Yet progress has been painfully slow in bringing sites to closure. Out of more than seventeen hundred sites that have been put on the Superfund cleanup list over the years, only about four hundred have been cleaned up and delisted.[28] More than half of the original 406 Superfund sites identified in 1983 remain on the list today.[29] As simpler sites have been cleaned up, the more difficult and expensive sites remain. Many of these involve groundwater.

Groundwater contamination occurs at most Superfund sites.[30] From the late 1970s to the 1990s, groundwater remediation (cleanup) projects relied on a relatively straightforward approach—pump the contaminated groundwater, treat it, and then inject it back into the aquifer. These pump and treat systems were intended to halt the spread of contamination *and* clean up the plume. In many cases, this technique proved effective in halting the spread of contamination but cleaning up the aquifer is another story altogether. Designers of early pump and treat systems assumed that by flushing a large amount of clean water through the aquifer, almost all of the contaminant would be removed. They were mistaken. After an extended period of pump and treat, it wasn't unusual to discover that the concentrations of contaminants rebounded after turning off the pumps. In the end, groundwater hydrologists began to realize that their pump and treat remedies were pulling a great deal of clean groundwater through aquifer material that was a highly contaminated source, creating newly contaminated groundwater that had to treated, without an endpoint in sight.

Part of the problem with pump and treat is the difficulty of removing contaminants attached (sorbed) onto aquifer surfaces. More critically, scientists began to appreciate the role of diffusion as a major problem with pump and treat systems. In the same way that a few drops of food coloring spread out in a glass of water, contaminants diffuse from higher to lower concentration areas in groundwater. Large amounts of contaminants can leave the main path of groundwater flow and become trapped in low-permeability and stagnant zones. Over time, the trapped contaminants slowly diffuse back to the main groundwater flow path, causing the contaminant to rebound when pump and treat stops. Fractured rocks are particularly challenging. The interconnected network of fractures provides the main pathway for groundwater flow, but diffusion can transfer a significant mass (often most) of the contaminant from the fractures into the rock matrix itself, which then becomes a long-term source of groundwater contamination.

Many of the most complex groundwater sites involve contamination by chlorinated solvents. Production of chlorinated solvents began in the United States in 1906 with carbon tetrachloride, followed by TCE and PCE in 1923. Widespread use began during World War II for degreasing metals in the electronic, instrument manufacturing, and aerospace industries, and increased markedly during the next three decades. At their peak, hundreds of millions of pounds of chlorinated solvents were produced each year.[31] For decades, chlorinated solvents were handled and disposed of haphazardly, often just being dumped into unlined pits, resulting in soil and groundwater contamination. The consequences were vividly portrayed in the book and major motion picture, *A Civil Action*, which chronicles the trial over childhood leukemia deaths linked to TCE-contaminated water wells in Woburn, Massachusetts.[32]

Dry cleaners also used large amounts of chlorinated solvents, particularly PCE. Wastewater containing these solvents was commonly poured down the drain or dumped on the ground behind the shop, contaminating groundwater and soils in thousands of locations across the country. The dry cleaner owners were primarily first-generation immigrants trying to make a living as best they could and weren't aware of the dangers.

The EPA has found TCE in more than one thousand of the seventeen hundred current or former Superfund sites.[33] Particularly troublesome is that when spilled in sufficient quantities, a portion can remain separate from water as a dense non-aqueous phase liquid, commonly referred to as a DNAPL (pronounced "D-nap-L"). Being denser than water, DNAPLs sink beneath the water table leaving residual contamination along the way as isolated "bull's-eyes" of contamination. If a DNAPL has enough mass, it can eventually pool on a clay or other low permeability layer in the subsurface. DNAPLs can be extraordinarily difficult to locate, let alone remove. They provide a long-term source of slowly dissolving contaminants to groundwater and can result in contaminant plumes several miles long. Even relatively small quantities of these chemicals can result in large areas of contamination exceeding drinking water standards.

As knowledge of the limitations of pump and treat and the complexities of DNAPLs slowly unfolded, many approaches were developed to supplement or replace pump and treat. Though far from easy, permeable reactive barriers were installed in deep trenches to treat groundwater as it flows through the barrier. Thermal techniques were developed to heat the ground to destroy or vaporize contaminants. These techniques come with large energy costs but are often effective. A common approach is to inject ingredients into the subsurface to make contaminants less harmful or to destroy them. One such approach, referred to as bioremediation, exploits microorganisms (microbes) to control and destroy contaminants. A key advantage of bioremediation and other in situ methods over pump and treat is that they eliminate the need for surface treatment and waste disposal.

Bioremediation is really just an extension of the work that microorganisms have done naturally for billions of years, breaking down human, animal, and plant wastes so that life on this planet can continue. Without microorganisms, the earth would literally be buried in wastes. Bioremediation works by simply adding more of the "right" microbes to eat the contaminants or by adding chemicals to stimulate microbial degradation. [34]

When Superfund was enacted, the state of knowledge concerning the degradation of organic compounds by microbes below the root zone of soils was roughly equivalent to the knowledge of the microbiology of the planet Mars. There were several reasons for this. Early studies of soils indicated that microbial populations dropped off sharply with depth. The concentration of natural organic materials in recharge also seemed too low to support life. Finally, groundwater was considered clean and wholesome because it was protected by soils. Consequently, there was no perceived need for such information. [35]

Recognition that groundwater was widely contaminated by organic pollutants dispelled the notion that it was protected by soils. In its place, the conventional wisdom became that, without expensive pump and treat, contaminated groundwater was irreversibly tainted. Once again, this thinking would be proven wrong. Observations that plumes resulting from gasoline spills often tended to shrink over time got microbiologists asking questions about the possible role of subsurface microbes in contaminant degradation. This issue was systematically investigated by a group of researchers brought together by the EPA and led by John T. Wilson, a microbiologist at the EPA Robert S. Kerr Environmental Research Center in Ada, Oklahoma. What they discovered is that bacteria in shallow aquifers are capable of degrading not just gasoline, but a variety of organic pollutants. This breakthrough led to an explosion of interest in subsurface microbiology.[36] However, as often occurs in science, this revolution in thinking took several twists and turns. The first assumption to be challenged concerned chlorinated solvents, which were widely considered to be resistant to biodegradation, particularly in oxygenated (aerobic) subsurface environments that occur in soils and many shallow aquifers. In 1983, John Wilson and his wife Barbara (an environmental chemist employed by the University of Oklahoma) demonstrated that TCE can be degraded in an aerobic environment by feeding methane to the microbes in a soil column. In metabolizing methane, the microbes produced an enzyme that, in turn, degraded TCE through a process known as cometabolism. Publication of the Wilsons' results received considerable attention.[37] Scientists at the EPA, Stanford University, and elsewhere began extensive lab and field investigations to see if there was a way to make this work in real-world plumes. Although progress was made, the degradation process could not be sustained. The investigations did, however, change thinking about the possibility of bioremediating organic pollutants in groundwater, igniting a flurry of research interest in this topic.

The first clear field demonstration of natural degradation of TCE in groundwater was at a Superfund site in St. Joseph, Michigan. The site, located about half a mile east of Lake Michigan, had been used for automobile brake manufacturing. For decades, workers had dumped TCE and other solvent wastes into unlined lagoons, resulting in massive soil and groundwater contamination. The groundwater contamination was first detected in 1982.

The EPA and Stanford researchers began to evaluate the potential for aerobic biodegradation of TCE by stimulating growth of those bacteria that metabolized methane. In one of those twists of happenchance that can occur in science, the researchers discovered that organic matter leaching from a disposal lagoon was depleting the oxygen and driving anaerobic (lacking oxygen) biodegradation. Amazingly, this natural rate of cleanup was faster than would be achieved by pump and treat or the planned bioremediation approach. This discovery suggested the possibility for manipulating conditions at contaminated sites to enhance anaerobic transformation of TCE to nontoxic end products.[38]

Having identified this unexpected process, the next (and more challenging) hurdle was to determine the rate and sustainability of that process. Wilson and his collaborators used detailed multi-level sampling combined with modeling to estimate the rates of biodegradation and the effects on contaminant fluxes through the aquifer. Since these early studies, an entire industry has been built around anaerobic biodegradation of chlorinated solvents by injecting emulsified vegetable oil, molasses, lactate, and other organic substrates. Bioremediation is now widely considered the default remedy for cleaning up chlorinated solvents in groundwater.[39]

Even with advances in bioremediation and similar remedies, a general problem persists—how do you get the remediation ingredients to where the contaminants are? High-resolution characterization of the geologic complexities of a contaminated site is a critical step to achieve this goal. Considerable progress has been made in these techniques during the past decade or so. Yet, time and again, taking shortcuts on this key task has led to clean up failures and delays. At a recent national remediation conference, 93 percent of attendees reported that failure to understand the geologic complexities of a site was the primary cause of performance shortcomings.[40] John Wilson makes an analogy to medicine. "If you look at scientific medicine as an example of people using science to do things right—diagnostics, lab tests, and the imaging is a major part of the bill. Sometimes it's more than half the cost of the cure. It's the cost of understanding what the problem is and what the best way is to cure it. In our business, we think we're spending too much if we spend more than 5 percent."[41] For those familiar with Tolkien's *Fellowship of the Ring*, Pippin's caution to Frodo summarizes the idea succinctly: "shortcuts make long delays."[42]

The EPA has a central role in groundwater contaminant cleanup, providing oversight to evaluate whether a project has a feasible design and whether ongoing projects are progressing toward successful completion. It's not just a matter of selecting a remediation technology and running with it. Different approaches may be phased in over time or for different parts of a plume. Care is also needed to avoid making conditions worse. For example, TCE can partially degrade to vinyl chloride—a more carcinogenic compound. There's also the very challenging problem of deciding what works and what doesn't—and under what circumstances. Remediation has become big business, accompanied with the standard problem of "snake-oil salesmen" selling products that are ineffective for the problem at hand.[43]

Finally, how clean is clean enough? Remediating an entire groundwater plume to drinking water standards is often a daunting, or just plain impossible, task. In many cases, the practical course for at least part of the plume is a transition to letting the natural system cleanse itself—an approach known as monitored natural attenuation. In order to gain public acceptance, this needs considerable evidence that it actually works. Periodic checks (monitoring) are required to see how things are going.

The EPA research center at Ada provided an exemplary example of how the EPA can meet these multi-faceted challenges of contaminant cleanup. While leading the research on bioremediation at the EPA, John Wilson spent a third to half of his time as an in-house consultant to assist people in selecting and implementing appropriate remediation technologies. His exposure to dealing with practical problems, in turn, helped identify key areas for further research— resulting in a combination of in-house expertise and collaboration with scientists in universities and other agencies.

The long arduous course of scientific study requires considerable time and patience—often in direct conflict to addressing the anxieties of a community affected by contamination. For Superfund (and other EPA programs) to be effective, the agency not only needs good scientists and lawyers, but also good communicators, listeners, and decision-makers with high ethical standards. To accomplish all this, the bottom line is that the EPA needs adequate funding and a favorable work environment to attract a capable and committed workforce.

12

A SUCCESS STORY

What the American public wants in the theater is a tragedy with
a happy ending.
—William Dean Howells [1]

The January 1963 issue of *Popular Science* featured the famous
rocket scientist Wernher von Braun. Buried inside this space-age
issue is some down to Earth advice on disposing of used engine oil:
"Dig a hole in the ground with a posthole digger and fill it with fine
gravel. Then pour in the oil. It will be absorbed into the ground
before your next change. Cover the spot with soil." [2] No mention
was made that used motor oil contains toxic chemicals and that a
little bit of oil can contaminate a lot of water. Nor were any cautions
given about not doing this near people's wells.

The *Popular Science* solution for getting rid of used oil is indica-
tive of the cavalier attitude given to disposal of all kinds of wastes
prior to formation of the U.S. Environmental Protection Agency
(EPA). Cities hauled municipal wastes to unlined dumps located on
the cheapest and most accessible land. Hazardous industrial wastes
were haphazardly mixed with the municipal wastes. Many businesses and rural residents took care of their own waste with out-of-
sight, out-of-mind disposal on vacant land or in the backyard. Trash
was often burned in the open, filling the air with smoke for miles.

"Midnight dumping" in rural areas or waterways was a common practice. Virtually no one considered the effects of toxic wastes seeping into groundwater. The gold standard during these years was a "sanitary landfill" that was merely covered with dirt to reduce sanitation hazards.[3]

The 1965 Solid Waste Disposal Act was the first attempt by the federal government to bring some order to solid waste practices, but it did little more than provide some basic guidelines for states to better control trash disposal. In 1971, EPA Administrator William Ruckelshaus initiated "Mission 5000" with the goal to close five thousand of the estimated fourteen thousand open dumps within a year. The deadline was not met, but it was a good first effort.[4] After passage of the Clean Air Act, the Clean Water Act, and the 1972 Ocean Dumping Act (which put a stop to offshore dumping of sewage sludge and industrial wastes), land disposal of wastes was the final regulatory frontier.

By the 1970s, hazardous wastes had become recognized as a serious and growing threat. A House committee warned that, "approximately 30–35 million tons of hazardous waste are literally dumped on the ground each year. Many of these substances can blind, cripple, or kill. They can defoliate the environment, contaminate drinking water supplies, and enter the food chain under present, largely unregulated disposal practices."[5]

In 1976, Congress passed the Resource Conservation and Recovery Act (RCRA). The act banned open dumping of solid waste and authorized the EPA to set minimum criteria for landfills. Hazardous wastes received special attention with strict requirements for their treatment, storage, and disposal. Businesses and industries also were tasked to keep "cradle-to-grave" records from the time hazardous waste is generated to its final disposal.

Solvents, battery acid, chemical wastes, and some pharmaceutical wastes are all examples of hazardous waste. In determining what's hazardous and what is not, the EPA addresses the following questions: Is it easily combustible or flammable? Is it corrosive or does it dissolve metals or burn the skin? Does it undergo a rapid or violent chemical reaction with other materials? Does it contain toxic metals, pesticides, or organic chemicals above regulatory levels? A yes answer to any one of these questions qualifies the waste as hazardous. The EPA also can target a waste as hazardous for other specific reasons.

Generators of hazardous waste include the full gamut, from large industries to small businesses, hospitals, universities, and government facilities. The RCRA cradle-to-grave regulations apply to sites generating more than one hundred kilograms (220 pounds) of hazardous waste per month (about half of a fifty-five-gallon drum) or more than one kilogram (2.2 pounds) a month of *acute* hazardous waste. Businesses that generate less than these amounts are still responsible for delivering their hazardous waste to an authorized site for storage, treatment, or disposal. This is not always easy. Complying with RCRA regulations can be very challenging for small businesses.

RCRA is one of the EPA's most important, but least appreciated, environmental laws. Elements of the program operate in nearly every community across the country. A total of 80 percent of all U.S. residents live within three miles of a RCRA-regulated hazardous waste generator or treatment, storage, or disposal facility. Half of the country's residents live within one mile.[6] Most states (called primacy states) basically run the show. These states have implemented a hazardous waste management program that is at least as stringent as the federal requirements. The EPA takes the lead for Tribal lands and non-primacy states. The RCRA program is reviewed annually by the EPA, but the policing of individual facilities is done by the states (when they have primacy) or initiated by citizen lawsuits.

As the title suggests, the RCRA also encourages practices that minimize waste generation and promote recycling. In this respect, the act has been a large success. The amount of hazardous waste generated in 2002 was about one-seventh of the amount generated annually when RCRA was enacted twenty-five years earlier.[7] In 2010, the Aspen Institute identified "Rethinking Wastes as Materials" as one of the top ten ways that the EPA has strengthened America.[8] An example of a major change in the use of hazardous materials is the switch from chlorinated solvents to ultrapure water for cleaning computer chips. Another RCRA hallmark has been its responsiveness to emerging and unique wastes—fluorescent light bulbs, lead acid batteries, medical wastes, and E-waste from computers and electronic gadgets—to ensure that these things don't end up just being thrown in the trash. And thanks to the 1980 Used Oil Recycling Act, almost all used oil is now recycled.

In 1980, nearly sixty thousand businesses treated, stored, or disposed of hazardous waste, with outdated practices resulting in contamination of many of these sites. The number of sites needing cleanup was more than triple those on the national Superfund list at the time.[9] In response, 1984 amendments created the RCRA Corrective Action Program, which required cleanup of all waste that had leaked into the environment at a hazardous waste facility. This cleanup, at the owner's expense, prevents expenditure of taxpayer dollars at a future Superfund site. Although cleanup at RCRA sites receives much less attention than Superfund, it is no less important. It's been a slow go. More progress has been made through a separate program under RCRA—the underground storage tank (UST) program.

As suburban areas rapidly expanded after World War II, oil companies bought prime real estate, built gas stations, and leased them to dealers. By 1971, the United States had around 225,000 gas stations, with most of them storing their fuel in underground tanks. At the same time, wells were being drilled to supply water to the fast-growing suburbs.[10]

By the 1980s, about two million USTs in the United States were located beneath gas stations, airports, military bases, schools, car rental agencies, and other businesses. Some tanks contained liquids such as pesticides, chemical wastes, and dry cleaning fluid, but most of them held gasoline and other petroleum fuels. Almost all USTs had been constructed of ordinary steel, making them highly susceptible to corrosion if they came into contact with groundwater. Assuming they were properly installed and maintained, these USTs had an estimated lifespan of fifteen to twenty years. At the time the problem got on the EPA's radar, it was estimated that at least a million of these steel petroleum storage tanks had been in the ground for more than sixteen years.[11]

Groundwater contaminated by gasoline can expose humans to numerous chemicals, including benzene, a proven human carcinogen. While gasoline has a familiar smell that often serves as a warning, low levels can be ingested in drinking water for an extended period of time before being detected. Vapors that collect in buildings at high concentrations can pose a threat of explosion or fire, and at lower concentrations they increase the risk of cancer and other detrimental health effects. These dangers had barely, if at all, been considered in siting gasoline and other chemical storage tanks underground.

In 1984, *60 Minutes*, the most popular television show at the time, aired "Check Your Water" (produced by Patti Hassler). The segment featured homeowners whose wells had been contaminated by leaking USTs in the Canob Park neighborhood of Richmond, Rhode Island. A nearby Mobil gas station was the prime suspect, but there was also an Exxon station across the street. The first complaint of foul-tasting water dated back to 1968.[12] Since then, the contamination had spread to more than a dozen wells, forcing residents to buy bottled water for drinking and cooking. One family told the *60 Minutes* commentator how they packed their kids in the car several times a week and drove to a relative's house just to take a bath. Most troubling was the fear that some children had been harmed in utero when their mothers unknowingly consumed contaminated water. Word about the contamination had spread, and no one could sell their house to escape the problem.

This wasn't the first instance of contamination from leaking USTs, but the *60 Minutes* episode brought the problem to national attention. Harry Reasoner, the moderator, reported that the EPA believed that leaking USTs could be the major pollution problem of the 1980s. (Many states later reported leaking USTs as the primary source of groundwater contamination.) Reasoner warned that the tanks beneath the corner gasoline service station in *your* neighborhood could be a "time bomb" ready to explode. The irony here was that USTs had been the solution for reducing the dangers of fire and explosions from aboveground tanks.

A Mobil executive vice president interviewed for the *60 Minutes* segment assured the television audience that the tanks at the Mobil gas station were not leaking. He insisted that, except for the rare case of human errors, "we don't have any leaks in tanks." The Exxon spokesman thought maybe 2 percent of their tanks were leaking. Another *60 Minutes* guest, a petroleum industry consultant, estimated that two or three out of every ten gas stations in the country were leaking gasoline into the ground. All three underestimated the extent of the problem.

As the number of leaking underground tanks grew, communities across the country were overwhelmed in dealing with them—including Richmond, Rhode Island. When asked why the town hadn't closed the gas station in Canob Park, the president of the town council responded, "It was brought to our attention, loud and clear, that a company the size of Mobil had resources to successfully counteract anything that we might put up!"[13]

Despite rising public outcry, the EPA was handicapped in being able to provide assistance. Gasoline is not considered a waste, which means it wasn't subject to RCRA. Congress, however, soon got the message. As part of the 1984 amendments to RCRA, a regulatory program for USTs was added. (Underground tanks for home heating oil and farm motor fuels remained exempt from the law.)

The EPA was now able to set standards for UST design and installation, leak detection, and spill and overfill control. However, the agency still lacked authority to assist states and communities in cleaning up leaking underground petroleum tanks. Congress had specifically excluded petroleum products from Superfund, the EPA's principal clean-up program. In 1986, Congress once again got the message and authorized cleanup of UST petroleum leaks.

Ron Brand was the EPA's first UST program director. He recognized that success of the program depended on states, tank owners, and tank operators all working together. Tank owners ranged from deep-pocketed oil conglomerates to mom and pop gas stations. "We wanted to be sure that the regulations were practical," Brand said.[14] "Perhaps the best way to begin defining EPA's responsibilities is to say what the Regional Offices will not be doing. They won't run the UST program for the state. They won't dictate behavior at the state level. They won't second-guess individual state decisions."[15] What this all boiled down to was that the EPA had structured more flexibility into the UST regulatory program than perhaps any other federal environmental program.

The UST program was set up as a franchise operation (think McDonald's) where the owner (in this case, the states) operates relatively independently within a larger organization's (the EPA's) framework. The program is financed in large part by a 0.1 cent tax on each gallon of motor fuel sold nationwide. Known as the LUST (Leaking Underground Storage Tank) Trust Fund, it provides money for states to fund staff, take enforcement actions, and undertake cleanups when necessary. The states have considerable latitude in how they use their LUST funds. (There's also the added fun of telling family and friends that you're going to a "lust" meeting.)

When the EPA completed its UST regulations in 1988, it gave tank owners ten years to upgrade or replace their tanks to meet the new standards. Less time (five years or less, depending on the age of the tank) was given to upgrade leak detection from the old-fashioned periodic dip-stick checks to sensitive probes capable of detecting extremely small leaks.

There have been financial and technical challenges along the way. On the financial side, tank owners are required to have one million dollars of insurance to cover future leaks. This can be a significant burden on small businesses who may only be able to afford a policy with a large deductible. However, by collecting their own small tax on gasoline sales, many states have established programs to make this insurance affordable with reasonable deductibles. Another financial challenge is that Congress repeatedly tries to raid the LUST Trust Fund for other purposes.

On the technical side, the presence of methyl tert-butyl ether (MTBE), the fuel additive that replaced tetraethyl lead in gasoline for two decades, has complicated cleanup and increased the cost at many sites, particularly in California and New England. Diesel and biofuels have also presented new challenges in corrosion control. And because petroleum products are lighter than water, a separate phase can float on the water table that is difficult to recover. Fortunately, improved scientific understanding of natural biodegradation processes in the soil and groundwater has allowed decisions to be made about "how clean is clean." For example, California has allowed thousands of "low-threat" USTs to be closed even when groundwater contaminant concentrations exceed drinking water standards in some portion of the site. Sites are eligible if remediation has been attempted, the dissolved plume is shrinking, and the groundwater appears to have no future as a drinking water source. The basic idea is that technical and funding resources available for environmental restoration are limited, and higher priority exists for these resources elsewhere.

By working with communities to prevent, detect, and clean up tank leaks, the UST program is regarded as one of the EPA's most successful programs. By 2018, more than 1.8 million substandard USTs had been closed and 478,000 releases had been cleaned up. The EPA continues to regulate over half a million UST systems, with the rate of newly reported leaks greatly reduced.[16] In 2015, the EPA upgraded its UST regulations to help further prevent and detect UST releases. As an indication of the general acceptance of the UST program, these new regulations received almost no press.

What about the Canob Park neighborhood? The town received a new water supply paid for in large part by the two oil companies (without admitting culpability). In 2015, the Exxon site received a "No Further Action" letter from the state. As of 2018, the Mobil station still has residual contamination in bedrock and the site continues to be monitored.

13

RESURRECTING THE EPA

> The environment is a problem you must tend to everlastingly. It doesn't go away. It's not like putting out a fire or even building a highway. You can't do it, then brush your hands and say, "on to the next task."
> —William D. Ruckelshaus [1]

The U.S. Environmental Protection Agency's (EPA's) mandate to protect the environment and public health is an immense and complicated task. Scott Pruitt advocated simplifying this thorny problem with a "back-to-basics" approach to environmental protection, portraying this as a commonsense return to the agency's roots. In point of fact, never in the EPA's history has there been a time when anything was ever simple. From its earliest days, the agency has been in a constant battle against powerful corporate interests. As noted by a former Republican staffer, the Trump administration's return to basics is "a smokescreen to their real intention to restore the dependence of the United States energy system on fossil fuels." [2]

For the past decade, the public has been subjected to a constant drumroll of misinformation directed against the EPA. American's have been told time and again that the EPA is out of control with regulations. The truth is far more complex. As we have seen, establishing any new regulation must build a strong case and overcome intense lobbying and litigation from affected interests. The other message by those who oppose the EPA, and this one has strong nostalgic appeal, is that the states are much better suited for dealing with their environmental problems than some monstrous federal agency. This is true in many respects but overlooks the fact that if the states by themselves had been doing their job there never would have been a need for a federal environmental protection agency. Ideally, states are best suited for implementing regulations and managing their resources, but federal laws, oversight, and technical support are essential.

In 2010, to recognize the EPA's fortieth anniversary, a group of environmental leaders convened by the Aspen Institute developed a list of the agency's top accomplishments.[3] Among them were major reductions in water pollution by industries and wastewater treatment plants; 75 to 90 percent less pollution from cars in 2010 than their 1970s counterparts; removing lead from gasoline; major progress on addressing acid rain; and banning DDT, which brought the bald eagle and other birds back from the verge of extinction. Also noteworthy are the EPA's unsung accomplishments in implementing lesser-known Congressional acts. The 1984 Emergency Planning and Community Right-to-Know Act requires industrial reporting of toxic releases and helps communities plan for chemical emergencies. The act was passed after a toxic gas release in Bhopal, India, killed thirty-eight hundred people. The 1989 Oil Pollution Act created a trust fund financed by a tax on oil to clean up spills when the responsible party is incapable or unwilling to do so. It was passed after the Exxon Valdez spilled eleven million gallons of crude oil into Alaska's Prince William Sound. Energy Star, a collaborative program between the EPA and Department of Energy, has guided consumers to energy-efficient appliances and products for over twenty-five years. This popular program (targeted for elimination by the Trump administration) saves consumers more than thirty billion dollars a year.[4]

While these accomplishments are largely in the past—what we might term the agency's Glory Days—the need for a strong EPA continues. On any given day, any number of environmental problems hit the news: dead zones in the Chesapeake Bay and Gulf of Mexico, cancer-causing chemicals in a city's drinking water, beach closures from fecal bacteria, lead in drinking water at an elementary school, rural wells contaminated by agricultural chemicals, and methane and toxic chemicals from fracking operations. The imperative for a strong EPA is no less today than it was fifty years ago.

Without vocal public insistence, the EPA never would have been created. And without this critical public support, the agency never would have achieved the accomplishments that many Americans now take for granted. As a society in general, we have entered the dangerous terrain of indifference towards the environment. This problem reflects a host of deep-seated and troublesome trends in America—loss of respect for expertise and fact-based evidence, loss of confidence in America's institutions, and undermining the press as "fake news" when you don't like the message.

With all these external forces at play and low morale among career employees at the EPA, restoring the agency to its rightful place will not be easy. A place to start is new leadership dedicated to the agency's mission. Upon being re-instated as EPA administrator after Anne Gorsuch resigned, William Ruckelshaus found a cadre of beaten-down enforcement personnel. He decided to dramatize his commitment for an effective enforcement program at a large meeting of EPA employees. Ruckelshaus explained that initially he was concerned that EPA enforcement personnel would be pent up like "a bunch of tigers in the tank" ready to pounce as soon as the lid came off. "Well, I think we opened the tank all right," he told the crowd, "but based on what I see here the past few months, there may be more pussycats in the tank than tigers." Ruckelshaus's speech was followed by sustained applause. EPA employees felt that, once again, the agency's top managers had a commitment to building a credible and effective enforcement effort.[5]

Along with new leadership, it is essential to restore the scientific capabilities of the EPA and faith in its scientific integrity. Joseph Goffman, a former senior EPA official and now executive director of the Harvard Environmental & Energy Law Program summed up the imperative: "Faithfully followed, the rulemaking process is a stern taskmaster that demands intellectual honesty."[6] Reversing corporate capture of the EPA, eliminating efforts to control scientists and their outputs, and undoing the damage to science-based rulemaking are key priorities. The EPA also needs adequate funding and a favorable work environment to attract a capable and committed workforce.

A key component of rebuilding the public's confidence in the agency and support of its mission is to connect the EPA to people's daily lives. Nearly half of the EPA's budget goes to popular and vital state-level programs, of which most people are unaware. To turn the tide, better communication with the public on these and other contributions should be one of the EPA's highest priorities.

The EPA's dual role as cop and helper is a difficult balancing act. No matter how necessary and well-intentioned, regulations trigger compliance costs. EPA opponents virtually never use the word regulation without inserting "job-killing" in front of it. This has created a perception that the EPA is detached from the social and economic realities facing many Americans, opening the agency to repeated (and generally false) attacks that repealing regulations is essential to protecting jobs. To counter these attacks, the EPA needs to find ways to make regulations simpler and less burdensome. People quickly become frustrated when faced with too many nitpicky and time-consuming regulations, particularly at the small business level. Recordkeeping and reporting costs for documenting compliance can be high and should be simplified whenever possible. Here, electronic reporting and advanced monitoring technologies, such as sensors and satellite imagery, show some promise in simplifying compliance monitoring.[7] Simpler and clearer permitting processes are also needed. This doesn't mean, of course, that the EPA should not enforce the rules it makes and ensure that companies conform to them.

Regulations are much easier to impose on new facilities and equipment than retroactively on those that already exist. For example, consider a water well contractor whose livelihood depends on an expensive drilling rig that he manages to keep running but that emits more air pollutants than allowable under today's standards for new drill rigs. Given the potential impacts on his livelihood, the better course would be to provide an incentive or grace period to upgrade, rather than a sudden demand to do so.

A rebooting of the EPA does not equate to simply a return to more of the 1970s-type approaches to environmental regulations. Rather it provides an opportunity to incorporate lessons learned about how to better design environmental policies through such means as market-based incentives, public-private partnerships, and a greater focus on encouraging processes and materials that minimize environmental pollution rather than just end of pipe treatment.[8] There's plenty of opportunity for more effective and user-friendly regulations when backed up by experiences and science. The solutions to many environmental issues also require greater integration across environmental media—air, surface water, groundwater, soil, and land. Collaborative efforts—such as the One Water movement to manage drinking water, wastewater, and storm water collectively to achieve multiple benefits—are a promising development.[9]

When an environmental crisis arises, public demands for action come fast and furious, whereas understanding the short- and long-term risks often takes time. In a true crisis, the EPA often has to make decisions on the fly based on the best available information and a delicate balancing act involving how much to apply the precautionary principle. The lead crisis in Flint, PFOA contamination of Hoosick Falls drinking water, and the shutdown of an entire public water supply system because of toxic algal blooms are situations when the EPA needs to be nimble and responsive to directing resources and attention to critical components of the crisis. In order to bolster public confidence and prevent an angry backlash when false assurances must be retracted, the agency needs to communicate risks—and uncertainties about those risks—in an understandable and forthright way.

EPA regulations will be effective in the long run only if they can withstand legal and legislative efforts to undo them. Obama learned that executive actions without Congressional support can be undone by following administrations. Michael Levi, a special assistant to

Obama on energy and economic policy had forewarned: "If you make a push purely on the executive action front and you don't back it up with measures to bolster public support, a lot of this can crumble under a new administration."[10] This is exactly what has happened. Fortunately, the Trump administration has fallen into the same trap with many of its actions.

A year into the Trump administration, an EPA employee summed up the situation at the beleaguered agency: "We're just kind of being told, 'Do the opposite thing you did 18 months ago.'"[11] In a post-Trump world that same advice applies. Addressing today's most pressing environmental problems requires the antitheses of Trump—cooperation, collective action, and a strong EPA. If you care about climate change, if you care about the environment, and if you care about your children's health—then you should care about the future of the U.S. Environmental Protection Agency.

NOTES

INTRODUCTION

1. S. Macdonald, *Propaganda and Information Warfare in the Twenty-First Century: Altered Images and Deception Operations* (New York: Taylor & Francis, 2007), 35.

I. EPA 101

1. C. S. Zarba, "The Assault Against Science Continues at the E.P.A.," *New York Times*, Nov. 14, 2018, https://www.nytimes.com/2018/11/14/opinion/environment-trump-epa-science.html.

2. E. G. Fitzsimmons, "Tap Water Ban for Toledo Residents," *New York Times*, Aug. 3, 2014, https://www.nytimes.com/2014/08/04/us/toledo-faces-second-day-of-water-ban.html.

3. T. Henry, "Water Crisis Grips Hundreds of Thousands in Toledo Area, State of Emergency Declared," *The Blade*, Aug. 3, 2014, https://www.toledoblade.com/local/2014/08/03/Water-crisis-grips-area.html.

4. M. A. Miller, et al., "Evidence for a Novel Marine Harmful Algal Bloom: Cyanotoxin (Microcystin) Transfer from Land to Sea Otters," *PLoS ONE* 5, no. 9 (2010): e12576.

5. J. L. Graham, N. M. Dubrovsky, and S. M. Eberts, *Cyanobacterial Harmful Algal Blooms and U.S. Geological Survey Science Capabilities* (Reston: U.S. Geological Survey, 2016), 2.

6. M. Cuevas, "Toxic Algae Bloom Blankets Florida Beaches, Prompts State of Emergency," *CNN*, July 1, 2016, https://www.cnn.com/2016/07/01/us/florida-algae-pollution/index.html.

7. P. Allitt, *A Climate of Crisis* (New York: Penguin, 2014), 65.

8. S. Samuel, "Lake Erie Just Won the Same Legal Rights as People," *Vox*, Feb. 26, 2019, https://www.vox.com/future-perfect/2019/2/26/18241904/lake-erie-legal-rights-personhood-nature-environment-toledo-ohio.

9. R. N. L. Andrews, "The EPA at 40: An Historical Perspective," *Duke Environmental Law & Policy Forum* (Spring 2011): 227.

10. M. K. Landy, M. J. Roberts, and S. R. Thomas, *The Environmental Protection Agency: Asking the Wrong Questions from Nixon to Clinton* (New York: Oxford University Press, 1994), 245.

11. W. A. Henry and G. Lee, "This Ice Queen Does Not Melt," *Time*, Jan. 18, 1982.

12. J. A. Mintz, *Enforcement at the EPA* (Austin: University of Texas Press, 1995), 41.

13. J. Lash, D. Sheridan, and K. Gillman, *A Season of Spoils: The Reagan Administration's Attack on the Environment* (New York: Pantheon, 1984), 45, 47; Mintz, *Enforcement at the EPA*, 49.

14. Henry and Lee, "This Ice Queen."

15. Mintz, *Enforcement at the EPA*, 50.

16. P. Shabecoff, "E.P.A. Chief Assailed on Lead Violation," *New York Times*, April 13, 1982, A18.

17. Landy, Roberts, and Thomas, *The Environmental Protection Agency*, 251.

18. M. E. Kraft, *Environmental Policy and Politics* (New York: Routledge, 2018), 140.

19. R. N. L. Andrews, *Managing the Environment, Managing Ourselves* (New Haven: Yale University Press, 2006), 432.

20. H. Doremus, "Scientific and Political Integrity in Environmental Policy," *Texas Law Review* 86 (2008): 1640.

21. T. Cama, "Senate GOP Steeling for Battle Against EPA," *The Hill*, Nov. 9, 2014, https://thehill.com/policy/energy-environment/e2-wire/223398-senate-gop-steeling-for-battle-against-the-epa.

22. D. Regas, "A Warning for Donald Trump: Gutting EPA Would Be Harder—And More Perilous—Than You Think," *Forbes*, Nov. 17, 2016, https://www.forbes.com/sites/edfenergyexchange/2016/11/17/a-warning-for-donald-trump-gutting-epa-would-be-harder-and-more-perilous-than-you-think/#4200d1954e80.

23. E. Schaeffer, "Reject Scott Pruitt for the E.P.A.," *New York Times*, Jan. 18, 2017, https://www.nytimes.com/2017/01/18/opinion/reject-scott-pruitt-for-the-epa.html.

24. C. Davenport, "Counseled by Industry, Not Staff, E.P.A. Chief Is Off to a Blazing Start," *New York Times*, July 1, 2017, https://www.nytimes.com/2017/07/01/us/politics/trump-epa-chief-pruitt-regulations-climate-change.html.

25. C. Davenport, "Scott Pruitt Is Seen Cutting the E.P.A. with a Scalpel, Not a Cleaver," *New York Times*, Feb. 5, 2017, https://www.nytimes.com/2017/02/05/us/politics/scott-pruitt-is-seen-cutting-the-epa-with-a-scalpel-not-a-cleaver.html.

26. C. Davenport, "Scott Pruitt, Trump's Rule-Cutting E.P.A. Chief, Plots His Political Future," *New York Times*, March 17, 2018, https://www.nytimes.com/2018/03/17/climate/scott-pruitt-political-ambitions.html.

27. F. Sonmez, et al., "From Charter Flights to Chick-fil-A: A Timeline of Scott Pruitt's Greatest Missteps," *Washington Post*, July 5, 2018, https://www.washingtonpost.com/politics/from-charter-flights-to-chick-fil-a-a-timeline-of-scott-pruitts-greatest-missteps/2018/07/05/1340f4ce-809c-11e8-b660-4d0f9f0351f1_story.html?utm_term=.c53ad472e467.

28. B. Dennis, and J. Eilperin, "Trump Plans to Nominate Andrew Wheeler, Former Coal Lobbyist, as EPA Chief," *Washington Post*, Nov. 16, 2018, https://www.washingtonpost.com/energy-environment/2018/11/16/trump-plans-nominate-andrew-wheeler-former-coal-lobbyist-permanent-epa-chief/?utm_term=.7155056e86cd.

29. Associated Press, "Court Orders Ban on Pesticide That Could Harm Babies' Brains, Says EPA Violated Law," *USA Today*, Aug. 10, 2018, https://www.usatoday.com/story/news/politics/2018/08/10/pesticide-ban-court-says-epa-violated-law-chlorpyrifos-harmful-babies/954921002/.

30. B. Dennis, J. Eilperin, and C. Mooney, "In Unprecedented Move, EPA to Block Scientists Who Get Agency Funding from Serving as Advisers," *Washington Post*, Oct. 30, 2017, https://www.washingtonpost.com/news/energy-environment/wp/2017/10/30/in-unprecedented-shift-epa-to-

prohibit-scientists-who-receive-agency-funding-from-serving-as-advisers/
?utm_term=.108a9543d635.

31. R. Showstack, "Environmental Ratings Lowest Ever for Congressional Republicans," *Eos* 99 (March 1, 2018), https://doi.org/10.1029/
2018EO094087.

32. Mintz, *Enforcement at the EPA*, 16.

33. J. Quarles, *Cleaning Up America: An Insider's View of the Environmental Protection Agency* (Boston: Houghton Mifflin, 1976), 51.

34. Ibid., 58–66.

35. U.S. Environmental Protection Agency, *William D. Ruckelshaus Oral History Interview* (Washington, DC: EPA Oral History Series, 1993), vii, 7.

2. TAKE IT FROM THE TAP

1. C. Hogue, "Rocket-Fueled River," *Chemical & Engineering News* 81, no. 33 (2003): 37–46.

2. U.S. Government Accountability Office, *Department of Defense Activities Related to Trichloroethylene, Perchlorate, and Other Emerging Contaminants* (Washington, DC: GAO-07-1042T, 2007), 9.

3. National Research Council, *Health Implications of Perchlorate Ingestion* (Washington, DC: National Academies Press, 2005).

4. Hogue, "Rocket-Fueled River."

5. The Kerr-McGee plant was the primary source of perchlorate contamination. A lesser source was a nearby perchlorate plant that had exploded in May 1988 after sparks from a welding torch started a fire. It was the largest domestic, non-nuclear explosion in recorded history.

6. G. Ayres, "A Little Rocket Fuel with Your Salad?" *World Watch Magazine* 16, no. 6 (2003): 12–20.

7. Hogue, "Rocket-Fueled River."

8. J. Eilperin, "A Frightening Map of Where Kerr-McGee Polluted," *Washington Post*, April 5, 2014, https://www.washingtonpost.com/news/
the-fix/wp/2014/04/05/where-did-kerr-mcgee-pollute-almost-every-state-
in-the-lower-48/?utm_term=.888736cc8696.

9. "Perchlorate," Nevada Division of Environmental Protection, accessed March 18, 2019, https://ndep.nv.gov/environmental-cleanup/site-
cleanup-program/active-cleanup-sites/bmi-complex/perchlorate.

10. Complicating matters, perchlorate is a naturally occurring impurity in nitrate salts imported from Chile's Atacama Desert to produce fertilizers. It also exists naturally in some soils at very low levels. When low levels of perchlorate are detected, debates may occur over who was responsible—NASA and the U.S. Department of Defense, or Mother Nature?

11. U.S. Government Accountability Office, *Perchlorate: Occurrence Is Widespread but at Varying Levels; Federal Agencies Have Taken Some Actions to Respond to and Lessen Releases* (Washington, DC: GAO-10-769, 2010).

12. Hogue, "Rocket-Fueled River."

13. A nationwide sampling of public water supplies completed in 2005 under the EPA Unregulated Contaminant Monitoring Rule detected perchlorate at or above four parts per billion in 160 of the 3,865 public water systems tested (about 4 percent). In thirty-one of these 160 systems, perchlorate was found above fifteen parts per billion.

14. D. Corn, "How a Clean Water Advocate and Senator Became a Chemical Industry Lobbyist," *Mother Jones*, Feb. 13, 2009, https://www.alternet.org/2009/02/how_a_clean_water_advocate_and_senator_became_a_chemical_industry_lobbyist/.

15. U.S. Government Accountability Office, *Perchlorate: Occurrence Is Widespread.*

16. A. Snider, "What Broke the Safe Drinking Water Act?" *Politico*, May 11, 2017, https://www.politico.com/agenda/story/2017/05/10/safe-drinking-water-perchlorate-000434.

17. J. Sapien, "Jackson to be Asked About Regulating Perchlorate in Drinking Water," *ProPublica*, Jan. 13, 2009, https://www.propublica.org/article/jackson-to-be-asked-about-regulating-perchlorate-in-drinking-water-090113.

18. U.S. Senate Committee on Environment and Public Works, "Hearing on the Nomination of Lisa P. Jackson to be Administrator of the U.S. Environmental Protection Agency and Nancy Helen Sutley to be Chairman of the Council on Environmental Quality," Senate Hearing 111-1178, Jan. 14, 2009, https://www.congress.gov/111/chrg/shrg94020/CHRG-111shrg94020.htm.

19. G. Nelson and E. Schor, "EPA Moves to Limit Perchlorate, Other Toxics," *Greenwire*, Feb. 2, 2011, https://www.eenews.net/stories/1059944757.

20. M. J. McGuire, *The Chlorine Revolution: Water Disinfection and the Fight to Save Lives* (Denver: American Water Works Association, 2013).

21. J. Salzman, *Drinking Water: A History* (New York: Overlook Duckworth, 2017).

22. "A Public Health Giant Step: Chlorination of U.S. Drinking Water," Water Quality & Health Council, May 1, 2008, http://www.waterandhealth.org/drinkingwater/chlorination_history.html.

23. D. Sedlak, *Water 4.0: The Past, Present, and Future of the World's Most Vital Resource* (New Haven: Yale University Press, 2014).

24. J. J. Rook, "The Formation of Halogens during Chlorination of Natural Waters," *Water Treatment and Examination* 23, no. 2 (1974): 234–43.

25. J. F. Pankow and J. A. Cherry, *Dense Chlorinated Solvents and other DNAPLs in Groundwater* (Portland: Waterloo Press, 1996).

26. H. M. Schmeck, "Water from Mississippi River Linked to Cancer Death Trends," *New York Times*, Nov. 8, 1974, 29.

27. H. M. Schmeck, "E.P.A. Orders a National Study of Chemical Contaminants in Drinking Water," *New York Times*, Nov. 9, 1974, 32.

28. Pankow and Cherry, *Dense Chlorinated Solvents*.

29. T. A. Bellar, J. J. Lichtenberg, and R. C. Kroner, "The Occurrence of Organohalides in Chlorinated Drinking Waters," *Journal of the American Water Works Association* 66, no. 12 (1974): 703–06.

30. J. M. Symons, et al., "National Organics Reconnaissance Survey for Halogenated Organics," *Journal of the American Water Works Association* 67, no. 11 (1975): 634–47.

31. V. J. Kimm, et al., "The Safe Drinking Water Act: The First 10 Years," *Journal of the American Water Works Association* 106, no. 8 (2014): 84–95.

32. Sedlak, *Water 4.0*, 97.

33. X.-F. Li and W. A. Mitch, "Drinking Water Disinfection Byproducts (DBPs) and Human Health Effects: Multidisciplinary Challenges and Opportunities," *Environmental Science & Technology* 52, no. 4 (2018): 1681.

34. Sedlak, *Water 4.0*, 103.

35. Li and Mitch, "Drinking Water Disinfection Byproducts."

36. Kimm, et al., "The Safe Drinking Water Act," 90.

37. Sedlak, *Water 4.0*, 107.

38. M. Messner, et al., "An Approach for Developing a National Estimate of Waterborne Disease Due to Drinking Water and a National Model Application," *Journal Water Health* 4 (2006): 201.

39. A. Thompkins, "Drinking Water Protection Program Update: Key Topics," Presentation at EPA National Drinking Water Advisory Council Meeting, Dec. 6, 2018.

40. M. Tiemann, *Drinking Water State Revolving Fund (DWSRF): Program Overview and Issues* (Washington, DC: Congressional Research Service, 2017).

41. U.S. Environmental Protection Agency, *Drinking Water Infrastructure Needs Survey and Assessment, Sixth Report to Congress* (Washington, DC: Office of Water, 2018).

42. The 1974 Safe Drinking Water Act also established rules for injection of liquid wastes to protect underground sources of drinking water and set up a process for special treatment of aquifers designated as sole-source aquifers.

43. C. A. Dieter, et al., *Estimated Use of Water in the United States in 2015* (Reston: U.S. Geological Survey, 2018).

44. The America's Water Infrastructure Act of 2018 allows funds from EPA's Drinking Water State Revolving Fund to be used for delineating and assessing source water protection areas for the first time since 1997.

45. Wellhead protection areas are determined by techniques ranging from mathematical groundwater models to a fixed radius around a well.

46. Pacific Institute, "Bottled Water and Energy: Getting to 17 Million Barrels," *Fact Sheet* (December 2007).

47. M. Allaire, H. Wu, and U. Lall, "National Trends in Drinking Water Quality Violations," *Proceedings, National Academy of Sciences* 115, no. 9 (2018): 2079.

3. ENVIRONMENTAL JUSTICE

1. M. Davey and M. Smith, "What Went Wrong in Flint," *New York Times*, Mar. 3, 2016, https://www.nytimes.com/interactive/2016/03/04/us/04flint-mistakes.html.

2. J. Salzman, *Drinking Water: A History* (New York: Overlook Duckworth, 2017), 151.

3. Ibid., 142.

4. Davey and Smith, "What Went Wrong."

5. A. Goodnough, M. Davey, and M. Smith, "When the Water Turned Brown," *New York Times*, Jan. 23, 2016, https://www.nytimes.com/2016/01/24/us/when-the-water-turned-brown.html.

6. Salzman, *Drinking Water*, 146.

7. D. Hohn, "Flint's Water Crisis and the 'Troublemaker' Scientist," *New York Times*, Aug. 8, 2016, https://www.nytimes.com/2016/08/21/magazine/flints-water-crisis-and-the-troublemaker-scientist.html.

8. Ibid.

9. M. Kennedy, "Lead-Laced Water in Flint: A Step-By-Step Look at the Makings of a Crisis," *NPR*, April 20, 2016, https://www.npr.org/sections/thetwo-way/2016/04/20/465545378/lead-laced-water-in-flint-a-step-by-step-look-at-the-makings-of-a-crisis.

10. Hohn, "Flint's Water Crisis."

11. Ibid.

12. Ibid.

13. Kennedy, "Lead-Laced Water."

14. Ibid.

15. M. Hanna-Attisha, *What the Eyes Don't See* (New York: One World, 2018).

16. Kennedy, "Lead-Laced Water."

17. Associated Press, "Doctors Urge Flint to Stop Using Water from Flint River," *Washington Times*, Sept. 24, 2015, https://www.washingtontimes.com/news/2015/sep/24/flint-plans-advisory-about-curbing-exposure-to-lea/.

18. Kennedy, "Lead-Laced Water."

19. C. Kozacek, "Flint's Contaminated Drinking Water Is Third Water Crisis for Michigan Governor," *Circle of Blue*, Jan. 11, 2016, https://www.circleofblue.org/2016/great-lakes/flints-contaminated-drinking-water-is-third-water-threat-for-michigan-governor/.

20. J. Lynch, "Michigan DEQ Vows Changes in Flint Water Crisis," *Detroit News*, Oct. 19, 2015, https://www.detroitnews.com/story/news/environment/2015/10/18/deq-mistakes/74198882/.

21. J. Lynch, "EPA Stayed Silent on Flint's Tainted Water," *Detroit News*, Jan. 12, 2016, https://www.detroitnews.com/story/news/politics/2016/01/12/epa-stayed-silent-flints-tainted-water/78719620/.

22. Salzman, *Drinking Water*, 148–49.

23. S. Atkinson and M. Davey, "5 Charged with Involuntary Manslaughter in Flint Water Crisis," *New York Times*, June 14, 2017, https://www.nytimes.com/2017/06/14/us/flint-water-crisis-manslaughter.html.

24. D. Shultz, "Was Flint's Deadly Legionnaires' Epidemic Caused by Low Chlorine Levels in the Water Supply?" *Science*, Feb. 5, 2018, https://www.sciencemag.org/news/2018/02/was-flint-s-deadly-legionnaires-epidemic-caused-low-chlorine-levels-water-supply.

25. J. Fortin, "Michigan Will No Longer Provide Free Bottled Water to Flint," *New York Times*, April 8, 2018, https://www.nytimes.com/2018/04/08/us/flint-water-bottles.html.

26. H. Gómez and K. Dietrich, "The Children of Flint Were Not 'Poisoned,'" *New York Times*, July 22, 2018, https://www.nytimes.com/2018/07/22/opinion/flint-lead-poisoning-water.html.

27. W. Finnegan, "Flint and the Long Struggle Against Lead Poisoning," *The New Yorker*, Feb. 4, 2016, https://www.newyorker.com/news/daily-comment/flint-and-the-long-struggle-against-lead-poisoning.

28. Atkinson and Davey, "5 Charged."

29. D. Kasler, P. Reese, and R. Sabalow, "360,000 Californians Have Unsafe Drinking Water. Are You One of Them?" *Sacramento Bee*, June 1, 2018, https://www.sacbee.com/news/state/california/water-and-drought/article211474679.html.

4. A WICKED PROBLEM

1. M. Warner, "Frank Perdue, 84, Chicken Merchant, Dies," *New York Times*, April 2, 2005, https://www.nytimes.com/2005/04/02/obituaries/frank-perdue-84-chicken-merchant-dies.html.

2. D. Shanker and L. Mulvany, "Perdue Unveils a More Humane Chicken Slaughter Process," *Bloomberg*, Feb. 26, 2019, https://www.bloomberg.com/news/articles/2019-02-26/perdue-unveils-a-more-humane-chicken-slaughter-process.

3. The EPA regulates CAFOs that exceed a certain number of animals (e.g., one thousand cattle, twenty-five hundred hogs, or 125,000 chickens), as well as smaller CAFOs that are identified as a significant contributor of pollutants.

4. M. Pollan, "Why Did the Obamas Fail to Take on Corporate Agriculture?" *New York Times*, Oct. 5, 2016, https://www.nytimes.com/

interactive/2016/10/09/magazine/obama-administration-big-food-policy.
html.

5. U.S. Environmental Protection Agency, "NPDES CAFO Regulations Implementation Status Reports," accessed May 2, 2019, https://www.
epa.gov/npdes/npdes-cafo-regulations-implementation-status-reports.

6. T. Fain, "Smithfield Rolls Out Biogas Plan, Says it Will Cover Most
Lagoons," *WRAL.com*, Oct. 25, 2018, https://www.wral.com/smithfield-
rolls-out-major-bio-gas-plan-lagoon-covers/17945911/.

7. D. Gurian-Sherman, *CAFOs Uncovered: The Untold Costs of Confined Animal Feeding Operations* (Cambridge: Union of Concerned Scientists, 2008).

8. M. Heller, "Will Storm Force a Hog Waste Reckoning in N.C.?"
Greenwire, Sept. 24, 2018, https://www.eenews.net/stories/1060099439.

9. R. Smothers, "Spill Puts a Spotlight on a Powerful Industry," *New
York Times*, June 30, 1995, 10.

10. C. Copeland, *Animal Waste and Water Quality: EPA's Response to
the Waterkeeper Alliance Court Decision on Regulation of CAFOs* (Washington, DC: Congressional Research Service, 2011).

11. "NPDES CAFO Regulations Implementation."

12. B. Walton, "EPA Turns Away from CAFO Water Pollution," *Circle
of Blue*, Dec. 22, 2016, https://www.circleofblue.org/2016/water-policy-
politics/epa-turns-away-cafo-water-pollution/.

13. "California Dairy Quality Assurance Program: Compliance
Through Education," California Dairy Research Foundation, accessed May
2, 2019, http://cdrf.org/home/checkoff-investments/cdqap/.

14. J. Quarles, *Cleaning Up America: An Insider's View of the Environmental Protection Agency* (Boston: Houghton Mifflin, 1976), 155.

15. J. Hanlon, et al., "Water Quality: A Half-Century of Progress," EPA
Alumni Association, March 25, 2016, available at http://www.epaalumni.
org/hcp/.

16. E. W. Kenworthy, "Clean-Water Bill Is Law Despite President's
Veto," *New York Times*, Oct. 19, 1972, 26.

17. R. N. L. Andrews, *Managing the Environment, Managing Ourselves* (New Haven: Yale University Press, 2006), 237.

18. Hanlon, et al., "Water Quality: A Half-Century of Progress," 8.

19. C. Kozacek, "U.S. Clean Water Law Needs New Act for the 21st
Century," *Circle of Blue*, Aug. 20, 2015, https://www.circleofblue.org/
2015/world/u-s-clean-water-law-needs-new-act-for-the-21st-century/.

20. M. Perez, *Water Quality Targeting Success Stories* (Washington, DC: World Resources Institute and American Farmland Trust, 2017), v.

21. S. S. Batie, "Wicked Problems and Applied Economics," *American Journal of Agricultural Economics* 90, no. 5 (2008): 1176–91.

22. R. B. Alexander, et al., "Differences in Phosphorus and Nitrogen Delivery to the Gulf of Mexico from the Mississippi River Basin," *Environmental Science & Technology* 42, no. 3 (2008): 822.

23. D. Scavia, et al., "Ensemble Modeling Informs Hypoxia Management in the Northern Gulf of Mexico," *Proceedings of the National Academy of Sciences* 114, no. 33 (2017): 8823.

24. E. McLellan, et al., "Reducing Nitrogen Export from the Corn Belt to the Gulf of Mexico: Agricultural Strategies for Remediating Hypoxia," *Journal of the American Water Resources Association* 51, no. 1 (2015): 263–89.

25. C. Smith, "New Jersey-Size 'Dead Zone' Is Largest Ever in Gulf of Mexico," *National Geographic*, Aug. 2, 2017, https://news.nationalgeographic.com/2017/08/gulf-mexico-hypoxia-water-quality-dead-zone/.

26. "Minnesota's Legacy," accessed May 2, 2019, http://www.legacy.leg.mn/about-funds.

27. K. Potter, "Pressed by Angry Farmers, Republicans Target Dayton's Water Quality Law for Repeal," *Associated Press*, Jan. 13, 2017, https://www.twincities.com/2017/01/13/pressed-by-angry-farmers-republicans-target-daytons-water-quality-law-for-repeal/.

28. "Minnesota Agricultural Water Quality Certification Program," accessed Sept. 20, 2017, http://www.mda.state.mn.us/awqcp; D. Gunderson, "Random Acts of Conservation: Water Quality Depends on Farmers' Willingness, Not Regulation," *MPRnews*, May 17, 2016, https://www.mprnews.org/story/2016/05/17/water-buffalo-red-river-agriculture-erosion.

29. J. Marcotty, "Overuse of Farm Fertilizer Drives Minnesota's First Effort to Regulate It," *Star Tribune*, April 25, 2018, http://www.startribune.com/overuse-of-farm-fertilizer-drives-state-s-first-effort-to-regulate-it/480757011.

30. R. Steinzor and S. Jones, "Collaborating to Nowhere: The Imperative of Government Accountability for Restoring the Chesapeake Bay," *Journal of Energy & Environmental Law* (Winter 2013): 51–67.

31. "Chesapeake Bay TMDL Fact Sheet," U.S. Environmental Protection Agency, accessed May 2, 2019, https://www.epa.gov/chesapeake-bay-tmdl/chesapeake-bay-tmdl-fact-sheet.

32. "2016 State of the Bay Report," Chesapeake Bay Foundation, accessed May 2, 2019, http://www.cbf.org/about-the-bay/state-of-the-bay-report/2016/index.html.

33. D. Fears, "The Chesapeake Bay Hasn't Been This Healthy in 33 Years, Scientists Say," *Washington Post*, June 15, 2018, https://www. washingtonpost.com/news/energy-environment/wp/2018/06/15/the-chesapeake-bay-hasnt-been-this-healthy-in-33-years-scientists-say/?utm_ term=.de46a3f91017.

34. J. S. Lefcheck, et al., "Long-Term Nutrient Reductions Lead to the Unprecedented Recovery of a Temperate Coastal Region," *Proceedings of the National Academy of Sciences* 115, no. 14 (2018): 3658–62.

35. Fears, "The Chesapeake Bay Hasn't Been This Healthy."

36. J. E. Colburn, "Coercing Collaboration: The Chesapeake Bay Experience," *William & Mary Environmental Law and Policy Review* 40, no. 1 (2016): 678.

37. Natural Resources Conservation Service, *Chesapeake Bay Progress Report* (Washington, DC: U.S. Department of Agriculture, Sept. 2016), 1.

38. Steinzor and Jones, "Collaborating to Nowhere," 53.

39. K. Blankenship, "Midpoint Assessment for Bay Cleanup: Only 40% of Nitrogen Goal Met," *Bay Journal*, July 9, 2018, https://www. bayjournal.com/article/midpoint_assessment_for_bay_cleanup_only_40_ of_nitrogen_goal_met.

40. K. Blankenship, "PA Plan Says It Will Increase Ag Inspections, Plant More Trees," *Bay Journal*, Feb. 28, 2016, https://www.bayjournal. com/article/pa_plan_says_it_will_increase_ag_inspections_plant_more_ trees.

41. D. Morelli, "Inspectors Find Most PA Farms, While Not All in Compliance, Are Trying," *Bay Journal*, May 2, 2017, https://www. bayjournal.com/article/inspectors_find_most_pa_farms_while_not_in_ compliance_are_trying.

42. J. Portnoy, "Trump Wants to End Funding of the Chesapeake Bay Cleanup. Here's Who's Fighting Back," *Washington Post*, Mar. 18, 2017, https://www.washingtonpost.com/local/virginia-politics/trump-wants-to-end-funding-of-the-chesapeake-bay-cleanup-heres-whos-fighting-back/

2017/03/18/64a1f50c-0a5a-11e7-b77c-0047d15a24e0_story.html?utm_term=.b78c3a4a3e63.

43. K. Ross, "Iowa Cover Crop Acres Grow, but Rate Declines in 2017," *Des Moines Register*, March 21, 2017, https://www.desmoinesregister.com/story/opinion/editorials/2017/03/21/farmers-show-how-cover-crops-help-water-and-profits/99447186/.

44. C. S. Jones, et al., "Iowa Stream Nitrate and the Gulf of Mexico," *PLoS ONE* 13 no. 4 (2018): e0195930.

45. M. Elmer, "Des Moines Water Works Won't Appeal Lawsuit," *Des Moines Register*, April 11, 2017, https://www.desmoinesregister.com/story/news/2017/04/11/des-moines-water-works-not-appeal-lawsuit/100321222/.

46. D. L. Osmond, et al., "Farmers' Use of Nutrient Management: Lessons from Watershed Case Studies," *Journal of Environmental Quality* 44, no. 2 (2015): 382–90.

47. "One Water Hub," US Water Alliance, accessed May 2, 2019, http://uswateralliance.org/one-water.

48. E. Gies, "How the Growing 'One Water' Movement is Not Only Helping the Environment But Also Saving Millions of Dollars," *Ensia*, May 8, 2018, https://ensia.com/features/one-water/.

49. B. Walton, "2017 Preview: Window of Opportunity for U.S. Department of Agriculture," *Circle of Blue*, Jan. 5, 2017, https://www.circleofblue.org/2017/water-policy-politics/2017-preview-window-opportunity-u-s-department-agriculture/.

5. INCONVENIENT CONNECTIONS

1. B. Walton, "EPA Clean Water Rule Meets Political Pushback," *Circle of Blue*, Aug. 6, 2015, https://www.circleofblue.org/2015/world/epa-clean-water-rule-meets-political-pushback/.

2. C. A. Dieter, et al., *Estimated Use of Water in the United States in 2015* (Reston: U.S. Geological Survey, 2018).

3. W. M. Alley and R. Alley, *High and Dry: Meeting the Challenges of the World's Growing Dependence on Groundwater* (New Haven: Yale University Press, 2017), 87.

4. "Clean Water Act Coverage of 'Discharges of Pollutants' via a Direct Hydrologic Connection to Surface Water," *Federal Register* 83, no. 34 (Feb. 20, 2018): 7126–28.

5. Ibid.

6. A Wittenberg, "Republicans, States Alarmed About EPA Action on Groundwater," *E&E News*, April 19, 2018, https://www.eenews.net/stories/1060079509.

7. B. Walton, "U.S. Courts Issue Contradictory Rulings on Groundwater and the Clean Water Act," *Circle of Blue*, Feb. 7, 2018, https://www.circleofblue.org/2018/world/u-s-courts-issue-contradictory-rulings-groundwater-clean-water-act/.

8. N. C. Lauer, "Maui Case Leads Big Island to Mull Wastewater Discharge," *West Hawaii Today*, March 1, 2018, https://www.westhawaiitoday.com/2018/03/01/hawaii-news/maui-case-leads-big-island-to-mull-wastewater-discharge/.

9. E. Peterson, "Expert Says Herrington Lake Pollution Is Worse Than We Thought," *89.3 WFPL*, Nov. 17, 2017, https://wfpl.org/new-documents-suggest-pollution-in-herrington-lake-more-severe-than-previously-thought/.

10. Walton, "U.S. Courts Issue Contradictory Rulings."

11. K. Rushing, "Tribe Fought for Coal Ash Safeguards, Then Pruitt Came Along," *Earthjustice*, Oct. 31, 2017, https://earthjustice.org/blog/2017-october/tribe-fought-for-coal-ash-safeguards-then-pruitt-came-along.

12. M. Akers, "Moapa Tribal Leader Who Led Charge Against Reid Gardner Coal Plant Dies at 44," *Las Vegas Sun*, Jan. 31, 2018, https://lasvegassun.com/news/2018/jan/31/moapa-tribal-leader-who-fought-coal-plant-dies/.

13. K. Rogers, "Moapa Solar Plant Deal Signed," *Las Vegas Review-Journal*, Nov. 20, 2012, https://www.reviewjournal.com/business/energy/moapa-solar-plant-deal-signed/.

14. U.S. Environmental Protection Agency, "EPA Response to Kingston TVA Coal Ash Spill," accessed April 22, 2019, https://www.epa.gov/tn/epa-response-kingston-tva-coal-ash-spill.

15. "Coal Combustion Residuals Impoundment Assessment Reports," U.S. Environmental Protection Agency, accessed April 22, 2019, https://www.epa.gov/sites/production/files/2016-06/documents/ccr_impoundmnt_asesmnt_rprts.pdf.

16. D. Zucchino, "Duke Energy Fined $102 Million for Polluting Rivers with Coal Ash," *Los Angeles Times*, May 14, 2015, https://www.latimes.com/nation/la-na-duke-energy-coal-ash-20150514-story.html.

17. J. Thomas-Blate, "Middle Fork Vermilion—Under Siege a Second Time," *American Rivers*, May 21, 2018, https://www.americanrivers.org/2018/05/middle-fork-vermilion-under-siege-a-second-time/.

18. T. Crane, "Judge Dismisses Group's Federal Suit on Coal-Ash Pits on Jurisdictional Grounds," *Champaign/Urbana News-Gazette*, Nov. 16, 2018, http://www.news-gazette.com/news/local/2018-11-16/judge-dismisses-groups-federal-suit-coal-ash-pits-jurisdictional-grounds.html.

19. A. Reilly, "Court Sides with Greens, Finds Obama-Era Rule Weak," *Greenwire*, Aug. 21, 2018, https://www.eenews.net/stories/1060094865.

20. B. Dennis and J. Eilperin, "EPA Will Reconsider Obama-Era Safeguards on Coal Waste," *Washington Post*, Sept. 14, 2017, https://www.washingtonpost.com/news/energy-environment/wp/2017/09/14/epa-will-reconsider-obama-era-safeguards-on-coal-waste/?utm_term=.d6aa2f28e898.

21. J. Eilperin and B. Dennis, "EPA Eases Rules on How Coal Ash Waste Is Stored Across U.S.," *Washington Post*, July 17, 2018, https://www.washingtonpost.com/national/health-science/epa-eases-rules-on-how-coal-ash-waste-is-stored-across-the-us/2018/07/17/740e4b9a-89d3-11e8-85ae-511bc1146b0b_story.html?utm_term=.1b1279e5a6e1.

22. "Public Hearing: Proposed Rule on Hazardous and Solid Waste Management System: Disposal of Coal Combustion Residuals from Electric Utilities; Amendments to the National Minimum Criteria (Phase One)," U.S. Environmental Protection Agency, accessed April 3, 2019, https://www.epa.gov/coalash/forms/public-hearing-proposed-amendments-national-minimum-criteria.

23. Eilperin and Dennis, "EPA Eases Rules."

24. A. Wittenberg and E. M. Gilmer, "EPA Won't Regulate Pollution That Moves Through Groundwater," *Greenwire*, April 16, 2019, https://www.eenews.net/stories/1060169889.

25. T. E. Dahl, *Status and Trends of Wetlands in the Conterminous United States, 2004–2009* (Washington, DC: U.S. Fish and Wildlife Service, 2011).

26. J. D. Fretwell, J. S. Williams, and P. J. Redman, compilers, *National Water Summary on Wetland Resources* (Reston: U.S. Geological Survey, 1996), 3.

27. A. Wittenberg, "How George H.W. Bush (Eventually) Rescued U.S. Wetlands," *Greenwire*, Dec. 3, 2018, https://www.eenews.net/stories/1060108603.

28. O. A. Houck, "Hard Choices: The Analysis of Alternatives Under Section 404 of the Clean Water Act and Similar Environmental Laws," *University of Colorado Law Review* 60 (1989): 773–840.

29. J. Z. Cannon, *Environment in the Balance* (Cambridge: Harvard University Press, 2015), 180–85.

30. R. N. L. Andrews, *Managing the Environment, Managing Ourselves* (New Haven: Yale University Press, 2006), 377.

31. Cannon, *Environment in the Balance*, 186.

32. R. Waskom and D. J. Cooper, "Why Farmers and Ranchers Think the EPA Clean Water Rule Goes Too Far," *The Conversation*, Feb. 27, 2017, https://theconversation.com/why-farmers-and-ranchers-think-the-epa-clean-water-rule-goes-too-far-72787.

33. C. Kozacek, "U.S. Clean Water Law Needs New Act for the 21st Century," *Circle of Blue*, Aug. 20, 2015, https://www.circleofblue.org/2015/world/u-s-clean-water-law-needs-new-act-for-the-21st-century/.

34. U.S. Environmental Protection Agency, *Connectivity of Streams and Wetlands to Downstream Waters: A Review and Synthesis of the Scientific Evidence* (Washington, DC: Office of Research and Development, 2015).

35. L. Gatz, *Waters of the United States (WOTUS): Current Status of the 2015 Clean Water Rule* (Washington, DC: Congressional Research Service, 2018).

36. "What Are Prairie Potholes," American Rivers, accessed April 3, 2019, https://www.americanrivers.org/rivers/discover-your-river/prairie-potholes/.

37. A. Snider, "Obama Admin OK'd Controversial Rule Over Experts' Objections," *Greenwire*, July 27, 2015, https://www.eenews.net/stories/1060022487.

38. Waskom and Cooper, "Why Farmers and Ranchers."

39. Ibid.

40. A. Wittenberg, "Beaten in WOTUS Messaging War, Greens Gird for Trump Rule," *Greenwire*, July 30, 2018, https://www.eenews.net/stories/1060091615.

41. R. Lazarus, "Who's on First? District, Appeals Courts Grapple with Jurisdiction," *The Environmental Forum* 33, no. 5 (2016): 13.

42. J. P. Jacobs and A. Snider, "'Mr. Clean Water Act' Faces His Biggest Challenge," *Greenwire*, Sept. 30, 2015, https://www.eenews.net/stories/1060025570.

43. D. Kasler and R. Sabalow, "Settlement Reached in Federal Case of Modesto-Area Farmer Fined $2.8 Million for Plowing His Field," *Sacramento Bee*, August 15, 2017, https://www.sacbee.com/news/state/california/water-and-drought/article167296702.html.

44. M. Ye Hee Lee, "Trump's Claim That Waters of the United States Rule Cost 'Hundreds of Thousands' of Jobs," *Washington Post*, March 28, 2017, https://www.washingtonpost.com/news/fact-checker/wp/2017/03/02/trumps-claim-that-waters-of-the-united-states-rule-cost-hundreds-of-thousands-of-jobs/?utm_term=.02d3f7327738.

45. B. Walton, "Clean Water Rule Repeal Cannot Come at a Pen Stroke," *Circle of Blue*, March 3, 2017, https://www.circleofblue.org/2017/water-policy-politics/clean-water-rule-repeal-cannot-come-pen-stroke/.

46. P. Brasher, "Judge Revives Obama WOTUS Rule, Blocking Trump Suspension," *Agri-Pulse Communications*, Aug. 17, 2018.

47. A. Wittenberg, "Experts Predict Legal Trouble for Scalia-inspired Rule," *Greenwire*, Dec. 14, 2018, https://www.eenews.net/stories/1060109731.

6. A NEVER-ENDING BATTLE

1. J. McNally, "July 26, 1943: L.A. Gets First Big Smog," *Wired*, July 26, 2010, https://www.wired.com/2010/07/0726la-first-big-smog/.

2. B. Ross and S. Amter, *The Polluters* (New York: Oxford University Press, 2010), 75.

3. S. Gardner, "LA Smog: The Battle Against Air Pollution," *Marketplace*, July 14, 2014, https://www.marketplace.org/2014/07/14/sustainability/we-used-be-china/la-smog-battle-against-air-pollution.

4. Reductions of nitrogen oxides were also required by 1976.

5. Story of debates over delaying auto emission standards based on J. Quarles, *Cleaning Up America: An Insider's View of the Environmental Protection Agency* (Boston: Houghton Mifflin, 1976), 177–93.

6. R. Revesz and J. Lienke, *Struggling for Air: Power Plants and the "War on Coal"* (New York: Oxford University Press, 2016), 34, Kindle.

7. U.S. Environmental Protection Agency, *The Plain English Guide to the Clean Air Act* (Research Triangle Park: Office of Air Quality Planning and Standards, 2007), 9.

8. J. E. McCarthy and C. Copeland, *EPA Regulations: Too Much, Too Little, or On Track?* (Washington, DC: Congressional Research Service, 2016), 9.

9. Union of Concerned Scientists, "Fuel Economy and Emissions Standards for Cars and Trucks, Model Years 2017 to 2025," Fact Sheet, June 2016; Job estimation is in full-time equivalents.

10. U.S. Environmental Protection Agency, *The Plain English Guide*, 8.

11. J. Eilperin and B. Dennis, "EPA to Pull Back on Fuel-Efficiency Standards for Cars, Trucks in Future Model Years," *Washington Post*, March 3, 2017, https://www.washingtonpost.com/national/health-science/epa-to-pull-back-on-fuel-efficiency-standards-for-cars-trucks-in-future-model-years/2017/03/03/c4406b0c-0054-11e7-99b4-9e613afeb09f_story.html?utm_term=.d981ee84ac3a.

12. M. Daly and T. Krisher, "Trump Set to Roll Back Federal Fuel-Economy Requirements," *Associated Press*, March 7, 2017, https://www.voanews.com/a/trump-set-to-roll-back-federal-fuel-economy-requirements/3754818.html.

13. R. Read, "Sensing Trouble, Automakers Prepare to Bargain with EPA, CARB on Emissions," *Washington Post*, April 13, 2017, https://www.washingtonpost.com/cars/sensing-trouble-automakers-prepare-to-bargain-with-epa-carb-on-emissions/2017/04/13/38da20a6-2004-11e7-bb59-a74ccaf1d02f_story.html?utm_term=.92f0414a7fd7.

14. J. Eilperin and B. Dennis, "Trump Officials Prepare to Undo Fuel-efficiency Targets Despite Some Automakers' Misgivings," *Washington Post*, March 29, 2018, https://www.washingtonpost.com/national/health-science/trump-officials-prepare-to-undo-fuel-efficiency-targets-despite-some-automakers-misgivings/2018/03/29/d4043b74-32b0-11e8-8abc-22a366b72f2d_story.html?utm_term=.66c0a8adeb64.

15. R. Beene, "'Climate Change Is Real,' Carmakers Tell White House in Letter," *Bloomberg*, May 21, 2018, https://www.bloomberg.com/news/articles/2018-05-21/carmakers-tell-white-house-that-climate-change-is-real-in-letter.

16. H. Tabuchi, "Calling Car Pollution Standards 'Too High,' E.P.A. Sets Up Fight with California," *New York Times*, April 2, 2018, https://www.nytimes.com/2018/04/02/climate/trump-auto-emissions-rules.html.

17. D. Shepardson, "EPA Staff Disputed Claim Fuel Efficiency Plan Would Save Lives," *Reuters*, Aug. 14, 2018, https://www.reuters.com/article/us-autos-emissions/epa-staff-disputed-claim-fuel-efficiency-plan-would-save-lives-idUSKBN1KZ2CS.

18. C. Davenport, "Top Trump Officials Clash Over Plan to Let Cars Pollute More," *New York Times*, July 27, 2018, https://www.nytimes.com/2018/07/27/climate/trump-auto-pollution-rollback.html.

19. J. A. Dloughy, R. Beene, and J. Lippert, "EPA Chief Signals Showdown with California on Fuel Emission Standards," *Bloomberg*, March 13, 2018, https://www.bloomberg.com/news/articles/2018-03-13/epa-chief-signals-showdown-with-california-on-tailpipe-standards.

20. D. Kahn, "5 Things to Watch as the Trump Administration Weakens Car Rules," *Scientific American*, July 20, 2018, https://www.scientificamerican.com/article/5-things-to-watch-as-the-trump administration-weakens-car-rules/.

21. T. Puko and A. Leary, "Trump Administration Cuts Off Talks with California Over Fuel-Efficiency Standards," *Wall Street Journal*, Feb. 21, 2019, https://www.wsj.com/articles/trump-administration-cuts-off-talks-with-california-over-fuel-efficiency-standards-11550770073.

22. GM Corporate Newsroom, "General Motors Calls for National Zero Emissions Vehicle (NZEV) Program," Oct. 26, 2018, https://media.gm.com/media/us/en/gm/news.detail.html/content/Pages/news/us/en/2018/oct/1026-emissions.html.

23. A. M. Bento, et al., "Flawed Analyses of U.S. Auto Fuel Economy Standards," *Science* 362, no. 6419 (2018): 1119.

24. H. Tabuchi, "The Oil Industry's Covert Campaign to Rewrite American Car Emissions Rules," *New York Times*, Dec. 13, 2018, https://www.nytimes.com/2018/12/13/climate/cafe-emissions-rollback-oil-industry.html.

25. B. Berman, "EPA Chief Wheeler Says EV Standards Are 'Social Engineering,'" *InsideEVs*, Feb. 4, 2019, https://insideevs.com/epa-chief-ev-standards-emissions/.

26. P. Baker and C. Davenport, "Obama Orders New Efficiency for Big Trucks," *New York Times*, Feb. 18, 2014, https://www.nytimes.com/2014/02/19/us/politics/obama-to-request-new-rules-for-cutting-truck-pollution.html.

27. E. Lipton, "How $225,000 Can Help Secure a Pollution Loophole at Trump's E.P.A." *New York Times*, Feb. 15, 2018, https://www.nytimes. com/2018/02/15/us/politics/epa-pollution-loophole-glider-trucks.html.

28. T. Cama, "EPA Reverses Course on 'Super-Polluting' Truck Policy," *The Hill*, July 27, 2018, https://thehill.com/policy/energy-environment/399161-epa-reverses-course-on-super-polluting-truck-policy.

29. R. Beene, "Trump's EPA to Propose Tougher Big-Rig Pollution Rules, Source Says," *Bloomberg*, Nov. 12, 2018, https://www.bloomberg. com/news/articles/2018-11-12/trump-s-epa-is-said-to-propose-tougher-big-rig-pollution-rules.

7. COSTS, BENEFITS, AND POLITICS

1. G. Jacks, "Acid Precipitation," in *Regional Ground-Water Quality*, ed. W. M. Alley (New York: Wiley, 1993), 405.

2. N. Oreskes and E. M. Conway, *Merchants of Doubt* (New York: Bloomsbury Press, 2010), 67.

3. G. E. Likens and F. H. Bormann, "Acid Rain: A Serious Regional Environmental Problem," *Science* 184, no. 4142 (1974): 1176–79.

4. R. Revesz and J. Lienke, *Struggling for Air: Power Plants and the "War on Coal"* (New York: Oxford University Press, 2016), 85, Kindle.

5. D. Bolze and J. Beyea, "The Citizen's Acid Rain Monitoring Network," *Environmental Science & Technology* 23, no. 6 (1989): 645–46.

6. "Interstate Pollution: Smother My Neighbor," *The Economist*, Sept. 7, 2013, https://www.economist.com/united-states/2013/09/07/smother-my-neighbour.

7. T. Overton, "Supreme Court Revives CSAPR," *Power Magazine*, April 29, 2014, https://www.powermag.com/supreme-court-revives-csapr/.

8. J. Iman, "Cities with Most Air Pollution Revealed," *CNN*, April 22, 2016, https://www.cnn.com/2016/04/20/health/air-pollution-report-irpt/ index.html.

9. U.S. Environmental Protection Agency, *The Plain English Guide to the Clean Air Act* (Research Triangle Park: Office of Air Quality Planning and Standards, 2007), 1.

10. C. Mooney, "In a Surprising Study, Scientists Say Everyday Chemicals Now Rival Cars as a Source of Air Pollution," *Washington Post*, Feb. 15, 2018, https://www.washingtonpost.com/news/energy-environment/wp/

2018/02/15/in-a-surprising-study-scientists-say-everyday-chemicals-now-rival-cars-as-a-source-of-air-pollution/?utm_term=.1ec148a8305d.

11. U.S. Environmental Protection Agency, "Air Quality—National Summary," accessed April 4, 2019, https://www.epa.gov/air-trends/air-quality-national-summary; Particulates have multiple standards based on particle size and time frame (24-hour vs annual).

12. J. H. Cushman, Jr., "Clinton Sharply Tightens Air Pollution Regulations Despite Concern Over Costs," *New York Times*, June 26, 1997, A1.

13. The ozone standard applies to the annual fourth highest daily maximum eight-hour concentration averaged over three years. Ozone affects not only public health but also public welfare through its effects on forests and crop yields. As such, it has both primary (health-based) and secondary (welfare-based) standards. The two standards have often been the same.

14. J. M. Broder, "Re-Election Strategy Is Tied to a Shift on Smog," *New York Times*, Nov. 16, 2011, A1.

15. J. M. Broder, "Obama Administration Abandons Stricter Air-Quality Rules," *New York Times*, Sept. 2, 2011, A1.

16. Broder, "Re-Election Strategy."

17. Broder, "Obama Administration Abandons."

18. R. Walton, "EEI Took Middle Road, Pushed for 70 ppb Ozone Standard, Report Says," *Utility Dive*, Oct. 9, 2015, https://www.utilitydive.com/news/eei-took-middle-road-pushed-for-70-ppb-ozone-standard-report-says/407082/.

19. J. E. McCarthy and R. K. Lattanzio, *EPA's 2015 Ozone Air Quality Standards* (Washington, DC: Congressional Research Service, 2017), 21.

20. N. Kusnetz, "Scott Pruitt Plans to Radically Alter How Clean Air Standards Are Set," *InsideClimate News*, May 10, 2018, https://insideclimatenews.org/news/10052018/epa-clean-air-act-standards-health-data-smog-science-scott-pruitt-american-lung-association-naaqs.

21. McCarthy and Lattanzio, *EPA's 2015 Ozone*, 20–21.

22. S. Reilly, "EPA Delays Ozone Rule," *Greenwire*, June 6, 2017, https://www.eenews.net/stories/1060055629.

23. R. Valdmanis, "EPA Reverses Decision to Delay Smog Rule After Lawsuits," *Reuters*, Aug. 3, 2017, https://www.reuters.com/article/us-usa-epa-smog/epa-reverses-decision-to-delay-smog-rule-after-lawsuits-idUSKBN1AJ1V9.

24. V. Volcovici, "EPA Designates Areas Noncompliant with 2015 Ozone Standards," *Reuters*, May 1, 2018, https://www.reuters.com/article/

us-usa-epa-ozone/epa-designates-areas-non-compliant-with-2015-ozone-standards-idUSKBN1I244T.

25. "8-Hour Ozone (2015) Designated Area/State Information," U.S. Environmental Protection Agency, accessed April 4, 2019, https://www3. epa.gov/airquality/greenbook/jbtc.html.

26. G. T. Goldman and F. Dominici, "Don't Abandon Evidence and Process on Air Pollution Policy," *Science* 363, no. 6434 (2019): 1398–400.

27. J. E. McCarthy, *EPA Utility MACT: Will the Lights Go Out?* (Washington, DC: Congressional Research Service, 2012), 2.

28. Ibid., 2.

29. R. N. L. Andrews, *Managing the Environment, Managing Ourselves* (New Haven: Yale University Press, 2006), 374.

30. National Association of Manufacturers, "Manufacturers: Utility MACT Is Extremely Costly Regulation," *Press Release*, Dec. 21, 2011, https://www.nam.org/Newsroom/Press-Releases/2011/12/Manufacturers--Utility-MACT-Is-Extremely-Costly-Regulation/.

31. McCarthy, *EPA Utility MACT*, 1.

32. Ibid., 6.

33. R. Conniff, "Tuna's Declining Mercury Contamination Linked to U.S. Shift Away from Coal." *Scientific American*, Nov. 23, 2016, https://www.scientificamerican.com/article/tunas-declining-mercury-contamination-linked-to-u-s-shift-away-from-coal/.

34. S. Reilly, "In About-Face, Utilities Urge EPA to Keep Mercury Rule," *Greenwire*, July 11, 2018, https://www.eenews.net/stories/1060088801.

35. L. Friedman, "New E.P.A. Plan Could Free Coal Plants to Release More Mercury into the Air," *New York Times*, Dec. 28, 2018, https://www.nytimes.com/2018/12/28/climate/mercury-coal-pollution-regulations.html.

36. S. Patel, "Bipartisan Senators Urge EPA to Drop Proposed Changes to Mercury Rule," *POWER Magazine*, March 18, 2019, https://www.powermag.com/bipartisan-senators-urge-epa-to-drop-proposed-changes-to-mercury-rule/.

37. W. Cornwall, "Critics See Hidden Goal in EPA Data Access Rule," *Science* 360, no. 6388 (2018): 472–73.

38. Later studies using publicly available data corroborated these findings.

39. M. Healy, "Scientists Blast EPA Effort that Would Discredit Health Research in the Name of 'Transparency,'" *Los Angeles Times*, Aug. 24,

2018, https://www.latimes.com/science/sciencenow/la-sci-sn-epa-research-transparency-20180824-story.html.

40. Cornwall, "Critics See Hidden Goal," 472.

41. L. Friedman, "E.P.A. Announces a New Rule. One Likely Effect: Less Science in Policymaking," *New York Times*, Apr. 24, 2018, https://www.nytimes.com/2018/04/24/climate/epa-science-transparency-pruitt.html.

42. L. M. Jenkins, "US EPA's Science 'Transparency' Proposal Likely Delayed Until 2020," *Chemical Watch*, Oct. 25, 2018; https://chemicalwatch.com/71259/us-epas-science-transparency-proposal-likely-delayed-until-2020.

43. U.S. Environmental Protection Agency, "Progress Cleaning the Air and Improving People's Health," accessed April 4, 2019, https://www.epa.gov/clean-air-act-overview/progress-cleaning-air-and-improving-peoples-health.

44. U.S. Environmental Protection Agency, *The Benefits and Costs of the Clean Air Act from 1990 to 2020* (Washington, DC: Office of Air and Radiation, 2011).

45. J. G. Zivin and M. Neidell, "Air Pollution's Hidden Impacts," *Science* 359, no. 6371 (2018): 39–40.

46. C. Marziali, "LA Environmental Success Story: Cleaner Air, Healthier Kids," *USC News*, March 4, 2015, https://news.usc.edu/76761/las-environmental-success-story-cleaner-air-healthier-kids/.

47. Centers for Disease Control and Prevention, "Blood Lead Levels in Children Aged 1–5 Years—United States, 1999–2010," *Morbidity and Mortality Weekly Report* 62, no. 13 (2013): 245–48.

48. U.S. Environmental Protection Agency, *The Plain English Guide*, 18.

49. C. W. Tessum, et al., "Inequity in Consumption of Goods and Services Adds to Racial–Ethnic Disparities in Air Pollution Exposure," *Proceedings of the National Academy of Sciences* 116, no. 13 (2019): 6001–06.

8. CLIMATE CHANGE

1. M. Robinson, *Climate Justice* (New York: Bloomsbury, 2018).

2. "Bush Visits EPA Offices," *Washington Post*, Feb. 9, 1989, A17.

3. B. Dennis and J. Eilperin, "Trump Signs Order at the EPA to Dismantle Environmental Protections," *Washington Post*, March 28, 2017, https://www.washingtonpost.com/national/health-science/trump-signs-order-at-the-epa-to-dismantle-environmental-protections/2017/03/28/3ec30240-13e2-11e7-ada0-1489b735b3a3_story.html?utm_term=.24f8802a055b.

4. C. Sellers, "Trump and Pruitt are the Biggest Threat to the EPA in its 47 Years of Existence," *Vox*, July 1, 2017, https://www.vox.com/2017/7/1/15886420/pruitt-threat-epa.

5. Dennis and Eilperin, "Trump Signs Order."

6. C. Davenport and E. Lipton, "How G.O.P. Leaders Came to View Climate Change as Fake Science," *New York Times*, June 3, 2017, https://www.nytimes.com/2017/06/03/us/politics/republican-leaders-climate-change.html.

7. D. Matthews, "Donald Trump Has Tweeted Climate Change Skepticism 115 Times. Here's All of It," *Vox*, June 1, 2017, https://www.vox.com/policy-and-politics/2017/6/1/15726472/trump-tweets-global-warming-paris-climate-agreement.

8. "An America First Energy Plan," The White House, accessed April 21, 2019, https://www.heartland.org/_template-assets/documents/An%20America%20First%20Energy%20Plan.pdf.

9. D. Merica and R. Marsh, "Trump Budget Chief on Climate Change: 'We Consider That to be a Waste of Your Money,'" *CNN*, March 16, 2017, https://www.cnn.com/2017/03/16/politics/donald-trump-budget-cut-epa/index.html.

10. N. Jones, "How the World Passed a Carbon Threshold and Why It Matters," *Yale Environment 360*, Jan. 26, 2017, https://e360.yale.edu/features/how-the-world-passed-a-carbon-threshold-400ppm-and-why-it-matters.

11. Davenport and Lipton, "How G.O.P. Leaders."

12. P. J. Egan and M. Mullin, "Climate Change: US Public Opinion," *Annual Review of Political Science* 20 (2017): 209-27.

13. R. Bravender, "Obama Attorneys Confident as Legal 'Super Bowl' Kicks Off," *Greenwire*, Oct. 29, 2015, https://www.eenews.net/stories/1060027150.

14. J. Z. Cannon, *Environment in the Balance* (Cambridge: Harvard University Press, 2015), 60.

15. Ibid., 71.

16. Ibid., 61.

17. Ibid., 65.

18. J. Freeman and A. Vermeule, "Massachusetts v. EPA: From Politics to Expertise," Harvard Law School Program on Risk Regulation Research Paper No. 08-11, Aug. 2007, http://ssrn.com/abstract=1307811.

19. J. E. McCarthy et al., *EPA's Clean Power Plan for Existing Power Plants: Frequently Asked Questions* (Washington, DC: Congressional Research Service, 2017).

20. R. J. Revesz and J. Lienke, *Struggling for Air: Power Plants and the "War on Coal"* (New York: Oxford University Press, 2016), Kindle.

21. Ibid., 66.

22. Ibid., 67.

23. Ibid., 74.

24. R. N. L. Andrews, *Managing the Environment, Managing Ourselves* (New Haven: Yale University Press, 2006), 375.

25. Revesz and Lienke, *Struggling for Air*, 77.

26. B. Plumer, "How Obama's Clean Power Plan Actually Works—A Step-By-Step Guide," *Vox*, Aug. 5, 2015, https://www.vox.com/2015/8/4/9096903/clean-power-plan-explained.

27. McCarthy et al., *EPA's Clean Power Plan*, 12.

28. "Clean Power Plan," Competitive Enterprise Institute, accessed April 21, 2019, https://cei.org/cleanpowerplan.

29. McCarthy et al., *EPA's Clean Power Plan*, 33–34.

30. B. Plumer, "I Asked Legal Experts How Trump Could Kill Obama's Clean Power Plan. Here's What They Said," *Vox*, March 28, 2017, https://www.vox.com/energy-and-environment/2017/2/23/14691438/trump-repeal-clean-power-plan.

31. McCarthy et al., *EPA's Clean Power Plan*.

32. J. Delingpole, "Scott Pruitt Is Failing to Drain the Swamp at EPA," *Breitbart*, March 27, 2017, https://www.breitbart.com/politics/2017/03/27/delingpole-scott-pruitt-is-failing-to-drain-the-swamp-at-the-epa/.

33. M. James, "195 Countries Signed Paris Climate Agreement, 2 Oppose It. For Now," *USA Today*, May 31, 2017, https://www.usatoday.com/story/news/nation/2017/05/31/only-two-nations-out-197-oppose-climate-pact---and-us-may-next/102360164/.

34. C. Davenport, "Nations Approve Landmark Climate Accord in Paris," *New York Times*, Dec. 12, 2015, https://www.nytimes.com/2015/12/13/world/europe/climate-change-accord-paris.html.

35. Davenport and Lipton, "How G.O.P. Leaders."

36. L. Friedman, "Cost of New E.P.A. Coal Rules: Up to 1,400 More Deaths a Year," *New York Times*, Aug. 21, 2018, https://www.nytimes.com/2018/08/21/climate/epa-coal-pollution-deaths.html.

37. L. Friedman, "E.P.A. Will Ease Path to New Coal Plants," *New York Times*, Dec. 4, 2018, https://www.nytimes.com/2018/12/04/climate/epa-coal-carbon-capture.html.

38. J. A. Dlouhy, "Dire Climate Change Warnings Cut from Trump Power-Plant Proposal," *Bloomberg*, Sept. 4, 2018, https://www.bloomberg.com/news/articles/2018-09-04/dire-climate-change-warnings-cut-from-trump-power-plant-proposal.

39. D. Rice, "Earth Just Had its 400th Straight Warmer-Than-Average Month Thanks to Global Warming," *USA TODAY*, May 17, 2018, https://www.usatoday.com/story/news/world/2018/05/17/global-warming-april-400th-consecutive-warm-month/618484002/.

40. L. Friedman, "Cost of New E.P.A. Coal Rules: Up to 1,400 More Deaths a Year," *New York Times*, Aug. 21, 2018, https://www.nytimes.com/2018/08/21/climate/epa-coal-pollution-deaths.html.

41. S. DiSavino, "President Trump Can't Stop U.S. Coal Plants from Retiring," *Reuters*, Jan. 13, 2019, https://www.reuters.com/article/us-usa-trump-coal/president-trump-cant-stop-u-s-coal-plants-from-retiring-idUSKCN1P80BY.

42. V. Smil, "Trump's Coal Policy Will Likely Do Just What Obama's Did," *Washington Post*, March 29, 2017, https://www.washingtonpost.com/opinions/trumps-coal-policy-will-likely-do-just-what-obamas-did/2017/03/29/7c5bb868-14b4-11e7-9e4f-09aa75d3ec57_story.html?utm_term=.f6aeba0c080e.

43. Office of Management and Budget, *2017 Draft Report to Congress on the Benefits and Costs of Federal Regulations and Agency Compliance with the Unfunded Mandates Reform Act* (Washington, DC: Office of the President of the United States, 2018).

44. P. Horn, "U.S. Renewable Energy Jobs Employ 800,000+ People and Rising: in Charts," *InsideClimate News*, May 30, 2017, https://insideclimatenews.org/news/26052017/infographic-renewable-energy-jobs-worldwide-solar-wind-trump.

45. A. C. Revkin, "Obama's Second-Term Options on the Environment," *New York Times*, Jan. 19, 2013, https://dotearth.blogs.nytimes.com/2013/01/19/obamas-second-term-options-on-the-environment/.

46. "The EPA Is Rewriting the Most Important Number in Climate Economics," *The Economist*, Nov. 16, 2017, https://www.economist.com/united-states/2017/11/16/the-epa-is-rewriting-the-most-important-number-in-climate-economics.

47. Ibid.

48. B. Plumer, "Trump Put a Low Cost on Carbon Emissions. Here's Why It Matters," *New York Times*, Aug. 28, 2018, https://www.nytimes.com/2018/08/23/climate/social-cost-carbon.html.

49. National Academies of Sciences, Engineering, and Medicine, *Valuing Climate Damages: Updating Estimation of the Social Cost of Carbon Dioxide* (Washington, DC: The National Academies Press, 2017).

50. Plumer, "Trump Put a Low Cost."

51. T. Cohen, "State of the Union: Obama Calls for Action, With or Without Congress," *CNN*, Jan. 29, 2014, https://www.cnn.com/2014/01/28/politics/2014-state-of-the-union/index.html.

52. Climate Leadership Council, "Economists' Statement on Carbon Dividends," accessed April 21, 2019, https://www.clcouncil.org/economists-statement/.

53. T. Cama, "Poll Finds Support for Business-Backed Carbon Tax Plan," *The Hill*, Sept. 10, 2018, https://thehill.com/policy/energy-environment/405808-poll-finds-support-for-business-backed-carbon-tax-plan.

54. J. E. McCarthy and C. Copeland, *EPA Regulations: Too Much, Too Little, or On Track?* (Washington, DC: Congressional Research Service, 2016), 2.

55. F. Zakaria, "China Is Winning the Future. Here's How," *Washington Post*, Oct. 12, 2017, https://www.washingtonpost.com/opinions/china-is-winning-the-future-heres-how/2017/10/12/6af2a370-af87-11e7-9e58-e6288544af98_story.html?utm_term=.0ec28590e83e.

56. G. Vaidyanathan, "How Bad of a Greenhouse Gas Is Methane?" *Scientific American*, Dec. 22, 2015, https://www.scientificamerican.com/article/how-bad-of-a-greenhouse-gas-is-methane/.

57. C. Mooney and B. Dennis, "Obama Administration Announces Historic New Regulations for Methane Emissions from Oil and Gas," *Washington Post*, May 12, 2016, https://www.washingtonpost.com/news/energy-environment/wp/2016/05/12/obama-administration-announces-historic-new-regulations-for-methane-emissions-from-oil-and-gas/?utm_term=.da7b92dccf4b.

58. M. Daly, "Interior Moves to Delay Obama-Era Rule on Methane Emissions," *Associated Press*, Oct. 4, 2017, https://www.voanews.com/a/interior-moves-delay-obama-era-rule-methane-emissions/4056739.html.

59. Mooney and Dennis, "Obama Administration Announces."

60. J. Eilperin, S. Mufson, and P. Rucker, "The Oil and Gas Industry is Quickly Amassing Power in Trump's Washington," *Washington Post*, Dec. 14, 2016, https://www.washingtonpost.com/politics/the-oil-and-gas-industry-is-quickly-amassing-power-in-trumps-washington/2016/12/14/0d4b26e2-c21c-11e6-9578-0054287507db_story.html?utm_term=.b90d4f0e6858.

61. R. Marsh, "EPA Ordered to Enforce Obama-Era Methane Pollution Rule," *CNN*, Aug. 1, 2017, https://www.cnn.com/2017/07/31/politics/dc-circuit-epa-methane-rule/index.html.

62. "Senate Upholds Methane Control Rule," *Science* 356, no. 6339 (2017): 668.

63. D. Grandoni, "Congress Decided Against Repealing This Climate Rule. So the Trump Administration Is Undoing it," *Washington Post*, Oct. 4, 2017, https://www.washingtonpost.com/news/energy-environment/wp/2017/10/04/congress-decided-against-repealing-this-climate-rule-so-the-trump-administration-is-undoing-it/?utm_term=.8ffca05aa0d9.

64. J. Tollefson, "US EPA Proposes Weaker Methane Rule for Oil and Gas Industry," *Nature.com*, Sept. 11, 2018, https://www.nature.com/articles/d41586-018-06671-z.

65. R. A. Alvarez, et al., "Assessment of Methane Emissions from the U.S. Oil and Gas Supply Chains," *Science* 361, no. 6398 (2018): 186–88.

66. N. Oreskes and E. M. Conway, *Merchants of Doubt* (New York: Bloomsbury Press, 2010), 125.

67. A. Reilly, "In Hit to Obama Legacy, Court Rejects HFC Phaseout Effort," *Greenwire*, Aug. 8, 2017, https://www.eenews.net/stories/1060058529/print.

68. In 2018, Trump's EPA issued "guidance" that it would no longer apply any restrictions on the use of HFCs. A separate case challenging this guidance is currently (as of May 2019) being heard by the DC Circuit Court.

69. P. McKenna, "What's Keeping Trump from Ratifying a Climate Treaty Even Republicans Support?" *InsideClimate News*, Feb. 12, 2019, https://insideclimatenews.org/news/12022019/kigali-amendment-trump-

ratify-hfcs-short-lived-climate-pollutant-republican-business-support-montreal-protocol.

70. "Six of the G7 Commit to Climate Action. Trump Wouldn't Even Join Conversation," *InsideClimate News*, June 10, 2018, https://insideclimatenews.org/news/10062018/g7-summit-climate-change-communique-trump-allies-estranged-germany-france-canada.

71. I. Stanley-Becker, "Who Drew It? Trump Asks of Dire Climate Report, Appearing to Mistrust 91 Scientific Experts," *Washington Post*, Oct. 10, 2018, https://www.washingtonpost.com/news/morning-mix/wp/2018/10/10/who-drew-it-trump-asks-of-dire-climate-report-appearing-to-mistrust-91-scientific-experts/?utm_term=.84153696753c.

9. TOXIC CHEMICALS

1. A. S. Kolok, *Modern Poisons: A Brief Introduction to Contemporary Toxicology* (Washington, DC: Island Press, 2016).

2. M. Freemantle, *The Chemists' War: 1914–1918* (Cambridge, UK: Royal Society of Chemistry, 2014).

3. B. Ross and S. Amter, *The Polluters* (New York: Oxford University Press, 2010), 12.

4. Ibid.

5. Lead-arsenic pesticide story in Ross and Amter, *The Polluters*, 46–51.

6. Ross and Amter, *The Polluters*.

7. Ibid.

8. R. Foster, "Kepone: The 'Flour' Factory," *Richmondmag.com*, July 8, 2005, https://richmondmagazine.com/news/kepone-disaster-pesticide/.

9. E. Francis, "Conspiracy of Silence," *Sierra Magazine*, Sept./Oct. 1994, https://vault.sierraclub.org/sierra/200103/conspiracy.asp.

10. Ibid.

11. S. Jensen, "Report of a New Chemical Hazard," *New Scientist* 32 (1966): 612.

12. M.-M. Robin, *The World According to Monsanto* (New York: New Press, 2010), 25.

13. J. G. Koppe and J. Keys, "PCBs and the Precautionary Principle," in *The Precautionary Principle in the 20th Century: Late Lessons from Early Warnings*, ed. P. Harremoés, et al. (New York: Earthscan, 2002), 64–78.

14. R. W. Risebrough, et al., "Polychlorinated Biphenyls in the Global Ecosystem," *Nature* 220, no. 5172 (1968): 1098–1102.

15. S. Tanabe, "PCB Problems in the Future: Foresight from Current Knowledge," *Environmental Pollution* 50, no. 1-2 (1988): 5–28.

16. M.-M. Robin, *The World According to Monsanto* (New York: New Press, 2010), 25.

17. Koppe and Keys, "PCBs and the Precautionary Principle," 64.

18. Robin, *The World According to Monsanto*, 15.

19. Agency for Toxic Substances & Disease Registry, "Toxic Substances Portal—Polychlorinated Biphenyls (PCBs)," accessed April 7, 2019, https://www.atsdr.cdc.gov/ToxProfiles/tp.asp?id=142&tid=26.

20. Environmental Working Group, "Anniston, Alabama: Monsanto Knew About PCB Toxicity for Decades," accessed April 5, 2019, http://www.ewg.org/research/anniston-alabama/monsanto-knew-about-pcb-toxicity-decades.

21. C. Kozacek, "U.S. Clean Water Law Needs New Act for the 21st Century," *Circle of Blue*, Aug. 20, 2015, https://www.circleofblue.org/2015/world/u-s-clean-water-law-needs-new-act-for-the-21st-century/.

22. J.-P. Desforges, et al., "Predicting Global Killer Whale Population Collapse from PCB Pollution," *Science* 361, no. 6409 (2018): 1373–76.

23. "Hudson River Cleanup," U.S. Environmental Protection Agency, accessed April 5, 2019, https://www3.epa.gov/hudson/cleanup.html#quest1.

24. U.S. Government Accountability Office, *EPA Has Increased Efforts to Assess and Control Chemicals but Could Strengthen Its Approach* (Washington, DC: GAO-13-249, 2013), 1.

25. The TSCA also has been modified over time to address specific concerns, including asbestos mitigation in schools, monitoring and control of radon, and abatement of lead-based paint.

26. U.S. Government Accountability Office, *EPA Has Increased*, 13.

27. M. Kruse and R. Arrieta-Kenna, "The 7 Oddest Things Donald Trump Thinks," *POLITICO Magazine*, Oct. 13, 2016, https://www.politico.com/magazine/story/2016/10/the-7-oddest-things-donald-trump-thinks-214354.

28. R. A. Hites, "Dioxins: An Overview and History," *Environmental Science & Technology* 45, no. 1 (2011): 16–20.

29. Robin, *The World According to Monsanto*, 30–33.

30. Ibid.

31. B. E. Erickson, "Asbestos: Still a Global Menace," *Chemical & Engineering News* 94, no. 47 (2016): 28–31.

32. T. Dickinson, "The Eco-Warrior: Lisa Jackson's EPA," *Rolling Stone*, Jan. 20, 2010, https://www.rollingstone.com/politics/politics-news/the-eco-warrior-lisa-jacksons-epa-199050/.

33. J. M. Broder, "E.P.A. Chief Set to Leave; Term Fell Shy of Early Hope," *New York Times*, Dec. 27, 2012, https://www.nytimes.com/2012/12/28/science/earth/lisa-p-jackson-of-epa-to-step-down.html?smid=tw-nytimes&_r=1&.

34. Dickinson, "The Eco-Warrior."

35. Broder, "E.P.A. Chief Set to Leave."

36. Dickinson, "The Eco-Warrior."

37. C. Hogue, "Lisa P. Jackson," *Chemical & Engineering News* 88, no. 19 (2010): 14–18.

38. Broder, "E.P.A. Chief Set to Leave."

39. K. Ward, "DEP Inspectors Describe Early Scene at Freedom Leak Site," *Charleston Gazette-Mail*, Jan. 13, 2014, https://www.wvgazettemail.com/news/special_reports/dep-inspectors-describe-early-scene-at-freedom-leak-site/article_50247d9d-69f6-5533-9676-ff04532f6264.html.

40. Ibid.

41. L. Bernstein, "Chemical Spill into W.Va. River Spurs Closures, Run on Bottled Water," *Washington Post*, Jan. 10, 2014, https://www.washingtonpost.com/national/health-science/chemical-spill-into-wva-river-spurs-closures-run-on-bottled-water/2014/01/10/a6ec518a-7a0e-11e3-b1c5-739e63e9c9a7_story.html?utm_term=.52dcfc68a2be.

42. U.S. Chemical Safety and Hazard Investigation Board, *Chemical Spill Contaminates Public Water Supply in Charleston, West Virginia* (Washington, DC: Investigation Report 2014-01-I-WV, 2017).

43. M. Wines, "Owners of Chemical Firm Charged in Elk River Spill in WV," *New York Times*, Dec. 17, 2014, https://www.nytimes.com/2014/12/18/us/owners-of-chemical-company-charged-in-elk-river-spill.html.

44. K. Ward, "Ex-Freedom Official Southern Sentenced to 30 Days, $20K Fine," *Charleston Gazette-Mail*, Feb. 17, 2016, https://www.wvgazettemail.com/news/cops_and_courts/ex-freedom-official-southern-sentenced-to-days-k-fine/article_eea0f3fc-8094-5633-bd5e-7849e387d70a.html.

45. U.S. Chemical Safety and Hazard Investigation Board, *Chemical Spill Contaminates*.

46. D. Stegon, "Senators Call on Trump Administration for Smooth TSCA Transition," *Chemical Watch*, Dec. 1, 2016, https://chemicalwatch.com/51346/senators-call-on-trump-administration-for-smooth-tsca-transition.

47. A. C. Kaufman, "Trump's EPA Actually Seems to Be Doing A Pretty Good Job Regulating New Chemicals," *HuffPost*, June 6, 2017, https://www.huffingtonpost.com/entry/epa-new-chemicals_us_5936b0fbe4b013c4816aff71.

48. Ibid.

49. E. Lipton, "Why Has the E.P.A. Shifted on Toxic Chemicals? An Industry Insider Helps Call the Shots," *New York Times*, Oct. 21, 2017, https://www.nytimes.com/2017/10/21/us/trump-epa-chemicals-regulations.html.

50. A. Snider and A. Guillén, "EPA Staffers, Trump Official Clashed Over New Chemical Rules," *POLITICO*, June 22, 2017, https://www.politico.com/story/2017/06/22/trump-epa-energy-chemicals-clash-239875.

51. K. Franklin, "EPA Releases Updated TSCA Inventory," *Chemical Watch*, Feb. 20, 2019, https://chemicalwatch.com/74480/epa-releases-updated-tsca-inventory.

52. D. Grandoni and B. Dennis, "EPA Signals It Will Ban Toxic Chemical Found in Paint Strippers," *Washington Post*, May 10, 2018, https://www.washingtonpost.com/news/energy-environment/wp/2018/05/10/epa-signals-it-will-ban-toxic-chemical-found-in-paint-strippers/?utm_term=.0567191b80ea.

53. M. Dourson, "ACSH Explains: What's the Story on Methylene Chloride (DCM)?" *American Council on Science and Health*, June 26, 2018, https://www.acsh.org/news/2018/06/26/acsh-explains-whats-story-methylene-chloride-dcm-13122.

54. Given the military's widespread use of paint strippers on bases across the globe, the Department of Defense received a ten-year exemption on the grounds of national security.

55. J. Eilperin and B. Dennis, "EPA Bans Deadly Chemical Used in Paint Strippers—But Provides a Loophole for Commercial Operators," *Washington Post*, March 15, 2019, https://www.washingtonpost.com/climate-environment/2019/03/15/epa-bans-deadly-chemical-used-paint-strippers-provides-loophole-commercial-operators/?utm_term=.7da9cec2eddb.

56. S. Kaplan, "E.P.A. Delays Bans on Uses of Hazardous Chemicals," *New York Times*, Dec. 19, 2017, https://www.nytimes.com/2017/12/19/health/epa-toxic-chemicals.html.

57. Grandoni and Dennis, "EPA Signals."

58. S. Poticha, "Lowe's, Sherwin-Williams, Home Depot Do the Right Thing," *Natural Resources Defense Council Blog*, June 26, 2018, https://www.nrdc.org/experts/shelley-poticha/lowes-sherwin-williams-home-depot-do-right-thing.

59. J. Eilperin and B. Dennis, "EPA Bans Deadly Chemical Used in Paint Strippers—But Provides a Loophole for Commercial Operators," *Washington Post*, March 15, 2019, https://www.washingtonpost.com/climate-environment/2019/03/15/epa-bans-deadly-chemical-used-paint-strippers-provides-loophole-commercial-operators/?utm_term=.7da9cec2eddb.

60. Beyond the TSCA, Trump's EPA strove to weaken the Integrated Risk Information System, the agency's most scientifically rigorous source of information on chemical toxicity to humans and a common target of House Republicans: J. Daley, "Is the EPA Stifling Science on Chemical Toxicity Reports?" *Scientific American*, April 26, 2019, https://www.scientificamerican.com/article/is-the-epa-stifling-science-on-chemical-toxicity-reports/.

10. THE FOREVER CHEMICALS

1. D. Andrews and B. Walker, *Poisoned Legacy: Ten Years Later, Chemical Safety and Justice for DuPont's Teflon Victims Remain Elusive* (Washington, DC: Environmental Working Group, 2015), 6.

2. N. Rich, "The Lawyer Who Became DuPont's Worst Nightmare," *New York Times*, Jan. 6, 2016, https://www.nytimes.com/2016/01/10/magazine/the-lawyer-who-became-duponts-worst-nightmare.html.

3. Ibid.

4. Andrews and Walker, *Poisoned Legacy*, 7.

5. M. S. Reisch, "DuPont and Chemours Settle PFOA Suits," *Chemical & Engineering News*, Feb. 13, 2017, https://cen.acs.org/articles/95/web/2017/02/DuPont-Chemours-settle-PFOA-suits.html.

6. Rich, "The Lawyer Who Became."

7. M. Smith, "Miles from Flint, Residents Turn Off Taps in New Water Crisis," *New York Times*, Nov. 24, 2017, https://www.nytimes.com/2017/11/24/us/michigan-water-wolverine-contamination.html.

8. X. Lim, "Tainted Water: The Scientists Tracing Thousands of Fluorinated Chemicals in Our Environment," *Nature* 566, no. 7742 (2019): 26–29.

9. E. Lipton and J. Turkewitz, "E.P.A. Proposes Weaker Standards on Chemicals Contaminating Drinking Water," *New York Times*, April 25, 2019, https://www.nytimes.com/2019/04/25/us/epa-chemical-standards-water.html.

10. Lim, "Tainted Water."

11. D. Sedlak, "Fool Me Once," *Environmental Science & Technology* 50, no. 15 (2016): 7937–38.

12. P. Grandjean, "Delayed Discovery, Dissemination, and Decisions on Intervention in Environmental Health: A Case Study on Immunotoxicity of Perfluorinated Alkylate Substances," *Environmental Health* 17, no. 1 (2018): 62–67.

13. X. C. Hu, et al., "Detection of Poly- and Perfluoroalkyl Substances (PFASs) in U.S. Drinking Water Linked to Industrial Sites, Military Fire Training Areas, and Wastewater Treatment Plants, *Environmental Science & Technology Letters* 3, no. 10 (2016): 344–50.

14. A. Lustgarten, L. Song, and T. Buford, "Suppressed Study: The EPA Underestimated Dangers of Widespread Chemicals," *ProPublica*, June 20, 2018, https://www.propublica.org/article/suppressed-study-the-epa-underestimated-dangers-of-widespread-chemicals.

15. C. P. Higgins and J. A. Field, "Our Stainfree Future? A Virtual Issue on Poly- and Perfluoroalkyl Substances," *Environmental Science & Technology* 51, no. 11 (2017): 5859–60.

16. T. Frisch, "Small Towns in New York and Vermont Share a Water Contamination Crisis, But Not an Official Response," *In These Times*, Sept. 5, 2017, http://inthesetimes.com/rural-america/entry/20490/industrial-pollution-pfoa-drinking-water-contamination-epa-saint-gobain.

17. Timeline of events from August 2014 to March 2017 available at http://villageofhoosickfalls.com/Water/timeline.html, accessed Mach 21, 2019.

18. J. McKinley and V. Yee, "Water Pollution in Hoosick Prompts Action by New York State," *New York Times*, Jan. 27, 2016, https://www.

nytimes.com/2016/01/28/nyregion/new-york-testing-water-in-hoosick-falls-for-toxic-chemical.html.

19. J. McKinley, "Fears About Water Supply Grip Village That Made Teflon Products," *New York Times*, Feb. 28, 2016, https://www.nytimes.com/2016/02/29/nyregion/fears-about-water-supply-grip-village-that-made-teflon-products.html.

20. Ibid.

21. J. McKinley, "Pollutant Is Removed from Water in Hoosick Falls, NY, Cuomo Says," *New York Times*, March 13, 2016, https://www.nytimes.com/2016/03/14/nyregion/pollutant-is-removed-from-water-in-hoosick-falls-ny-cuomo-says.html.

22. C. Ward, "Numbers Released on Groundwater, Soil PFOA Contamination in Hoosick Falls," *News10.com*, June 19, 2017, https://www.news10.com/news/numbers-released-on-groundwater-soil-pfoa-contamination-in-hoosick-falls_20180313101614341/1037447026.

23. B. J. Lyons, "Hoosick Falls Added to Federal Superfund List," *Times Union*, July 31, 2017, https://www.timesunion.com/news/article/Hoosick-Falls-added-to-federal-Superfund-list-11720206.php.

24. J. Gullo, "NY Department of Health Issues Catch and Release Advisory for Anglers at Thayers Pond," *News10.com*, July 24, 2017, https://www.news10.com/news/ny-department-of-health-issues-catch-and-release-advisory-for-anglers-at-thayers-pond_20180313102638338/1037536494.

25. H. Bishop, "Hoosick Falls Still Waiting for Action on Water Problems," *Times Union*, June 6, 2017, https://www.timesunion.com/tuplus-opinion/article/Hoosick-Falls-still-waiting-for-action-on-water-11200390.php.

26. "EPA's PFAS Action Plan," U.S. Environmental Protection Agency, accessed April 6, 2019, https://www.epa.gov/pfas/epas-pfas-action-plan.

27. B. Dennis, "EPA Vows National Action on Toxic 'Forever' Chemicals," *Washington Post*, Feb. 14, 2019, https://www.washingtonpost.com/climate-environment/2019/02/14/epa-vows-national-action-toxic-forever-chemicals/?utm_term=.d924669bc5ee.

II. SUPERFUND

1. J. Lash, D. Sheridan, and K. Gillman, *A Season of Spoils: The Reagan Administration's Attack on the Environment* (New York: Pantheon, 1984), 112.

2. U.S. Government Accountability Office, *SUPERFUND: Trends in Federal Funding and Cleanup of EPA's Nonfederal National Priorities List Sites* (Washington, DC: GAO-15-812, 2015).

3. U.S. Environmental Protection Agency, "Superfund Enforcement: 35 Years of Protecting Communities and the Environment," accessed April 6, 2019, https://www.epa.gov/enforcement/superfund-enforcement-35-years-protecting-communities-and-environment.

4. National Research Council, *Alternatives for Managing the Nation's Complex Contaminated Groundwater Sites* (Washington, DC: National Academies Press, 2012), 1.

5. U.S. Department of Energy, "Environmental Cleanup," accessed April 6, 2019, https://energy.gov/national-security-safety/environmental-cleanup.

6. R. T. Nakamura and T. W. Church, *Taming Regulation: Superfund and the Challenge of Regulatory Reform* (Washington, DC: Brookings Institution, 2003), 10.

7. P. Shabecoff, "Rita Lavelle Gets 6-Month Term and Is Fined 10,000 for Perjury," *New York Times*, Jan. 10, 1984, A1.

8. M. K. Landy, M. J. Roberts, and S. R. Thomas, *The Environmental Protection Agency: Asking the Wrong Questions from Nixon to Clinton* (New York: Oxford University Press, 1994), 266.

9. B. Anderson, "Taxpayer Dollars Fund Most Oversight and Cleanup Costs at Superfund Sites," *Washington Post*, Sept. 20, 2017, https://www.washingtonpost.com/national/taxpayer-dollars-fund-most-oversight-and-cleanup-costs-at-superfund-sites/2017/09/20/aedcd426-8209-11e7-902a-2a9f2d808496_story.html?utm_term=.7319e37b36b8.

10. "Anaconda Closing Open-Pit Copper Mine at Butte," *New York Times*, April 25, 1982, 23.

11. G. Plaven, "Anaconda Smelter: 30 Years Later," *Montana Standard*, Sept. 26, 2010, https://mtstandard.com/news/local/anaconda-smelter-years-later/article_1177866e-c92d-11df-8251-001cc4c002e0.html.

12. N. Saks, "Butte, Anaconda Superfund Sites Added to EPA 'Emphasis List,'" *Montana Public Radio*, Dec. 15, 2017, https://www.mtpr.org/post/butte-anaconda-superfund-sites-added-epa-emphasis-list.

13. Associated Press, "Thousands of Snow Geese Die in Montana After Landing on Contaminated Water," *The Guardian*, Dec. 6, 2016, https://www.theguardian.com/us-news/2016/dec/07/thousands-of-snow-geese-die-in-montana-after-landing-on-contaminated-water.

14. N. Saks, "Drones, Lasers and Cannons: Hazing Birds from the Toxic Berkeley Pit," *Montana Public Radio*, Nov. 28, 2017, https://www.mtpr.org/post/drones-lasers-and-cannons-hazing-birds-toxic-berkeley-pit.

15. K. Sullivan and E. Hassler, "Residential Metals Abatement Program Offers Services for Free," *Montana Standard*, Apr. 28, 2016, https://mtstandard.com/natural-resources/superfund/residential-metals-abatement-program-offers-services-for-free/article_f015139f-cc3c-558b-bef7-01c1f719a64e.html.

16. "Natural Resource Damage Program," Montana Department of Justice, accessed March 17, 2019, https://dojmt.gov/lands/.

17. S. Dunlap, "The Cleanup: After 16 Years, Major Silver Bow Creek Remediation Nearly Complete," *Montana Standard*, March 30, 2015, https://mtstandard.com/news/local/the-cleanup-after-years-major-silver-bow-creek-remediation-nearly/article_b0400ec8-5c07-5fc9-ac04-758a51e647f0.html.

18. J. Robbins, "Montana Dam Is Breached, Slowly, to Restore a Superfund Site," *New York Times*, May 27, 2008, https://www.nytimes.com/2008/05/27/science/27dam.html.

19. S. Dunlap, "EPA to Start New Chapter of Butte Superfund in Spring," *Montana Standard*, March 9, 2019, https://mtstandard.com/news/local/epa-to-start-new-chapter-of-butte-superfund-in-spring/article_8c324dc4-9b5e-556b-a4b1-f0b9125fdbfe.html.

20. B. Tyer, *Opportunity, Montana: Big Copper, Bad Water, and the Burial of an American Landscape* (Boston: Beacon, 2014).

21. U.S. Environmental Protection Agency, "San Gabriel Valley Groundwater Cleanup Superfund Progress Report," May 2017, http://sgvog.org/_assets/2017%20San%20Gabriel%20Valley%20Groundwater%20Cleanup%20Progress%20Report%20FINAL%20%20Updates%205-17-17.pdf.

22. S. Scauzillo, "Contaminated Ground Water in San Gabriel Valley Gets $250 Million Boost, Extending Cleanup Until 2027," *San Gabriel*

Valley Tribune, June 4, 2017, https://www.sgvtribune.com/2017/06/04/contaminated-ground-water-in-san-gabriel-valley-gets-250-million-boost-extending-cleanup-until-2027/.

23. W. M. Alley and R. Alley, *High and Dry: Meeting the Challenges of the World's Growing Dependence on Groundwater* (New Haven: Yale University Press, 2017), 186–87.

24. U.S. Environmental Protection Agency, "Anatomy of Brownfields Redevelopment," accessed April 6, 2019, https://www.epa.gov/brownfields/anatomy-brownfields-redevelopment.

25. U.S. Environmental Protection Agency, "Brownfields Program Accomplishments and Benefits," accessed April 6, 2019, https://www.epa.gov/brownfields/brownfields-program-accomplishments-and-benefits.

26. National Environmental Justice Advisory Council, *Unintended Impacts of Redevelopment and Revitalization Efforts in Five Environmental Justice Communities* (Washington, DC: U.S. Environmental Protection Agency, 2006).

27. U.S. Environmental Protection Agency, *Addressing Environmental Justice in EPA Brownfields Communities* (Washington DC: Solid Waste and Emergency Response, 2009).

28. U.S. Environmental Protection Agency, "Deleted National Priorities List (NPL) Sites - by State," accessed April 7, 2019, https://www.epa.gov/superfund/deleted-national-priorities-list-npl-sites-state.

29. Anderson, "Taxpayer Dollars."

30. U.S. Environmental Protection Agency, *Superfund Remedy Report 15th Edition* (Washington, DC: 2017), 23.

31. J. F. Pankow and J. A. Cherry, *Dense Chlorinated Solvents and Other DNAPLs in Groundwater* (Portland, OR: Waterloo Press, 1996).

32. J. Harr, *A Civil Action* (New York: Random House, 1996).

33. Agency for Toxic Substances & Disease Registry, "Toxic Substances Portal—Trichloroethylene (TCE)," accessed April 7, 2019, https://www.atsdr.cdc.gov/phs/phs.asp?id=171&tid=30.

34. National Research Council, *In Situ Bioremediation* (Washington, DC: National Academy Press, 1993), 16.

35. J. T. Wilson, et al., "Enumeration and Characterization of Bacteria Indigenous to a Shallow Water-Table Aquifer," *Ground Water* 21, no. 2 (1983): 134–42.

36. F. H. Chapelle, *Ground-Water Microbiology and Geochemistry* (New York: John Wiley, 1993), 192.

37. J. T. Wilson and B. H. Wilson, "Biotransformation of Trichloroethylene in Soil," *Applied and Environmental Microbiology* 49, no. 1 (1985): 242–43.

38. L. Semprini, et al., "Anaerobic Transformation of Chlorinated Aliphatic Hydrocarbons in a Sand Aquifer Based on Spatial Chemical Distributions," *Water Resources Research* 31, no. 4 (1995): 1051–62.

39. T. H. Wiedemeier, et al., *Technical Protocol for Evaluating Natural Attenuation of Chlorinated Solvents in Ground Water* (Washington, DC: U.S. Environmental Protection Agency, 1998).

40. Conference was National Ground Water Association, "Groundwater Solutions: Innovating to Address Emerging Issues for Groundwater Resources," Arlington, VA, Aug. 8–9, 2017.

41. John Wilson, personal communication, June 2017.

42. J. R. R. Tolkien, *The Fellowship of the Ring* (London: George Allen & Unwin, 1954).

43. National Research Council, *In Situ Bioremediation*, 2.

12. A SUCCESS STORY

1. Howells cited in E. Wharton, *A Backward Glance* (New York: D. Appleton-Century, 1934), 147.

2. "Hints from the Model Garage," *Popular Science* (Jan. 1963), 166.

3. R. N. L. Andrews, *Managing the Environment, Managing Ourselves* (New Haven: Yale University Press, 2006), 245–47.

4. Ibid., 247.

5. U.S. Environmental Protection Agency, *25 Years of RCRA: Building on Our Past to Protect Our Future* (Washington, DC: Office of Solid Waste and Emergency Response, 2002), 2.

6. U.S. Environmental Protection Agency, *RCRA's Critical Mission & the Path Forward* (Washington, DC: EPA530-R-14-002, 2014), 10.

7. U.S. Environmental Protection Agency, *25 Years of RCRA*, 15.

8. Aspen Institute, *EPA 40th Anniversary: 10 Ways EPA Has Strengthened America* (Aspen: Energy and Environment Program, 2010).

9. U.S. Environmental Protection Agency, *25 Years of RCRA*, 6.

10. H. E. Brieger, "Lust and the Common Law: A Marriage of Necessity," *Boston College Environmental Affairs Law Review* 13, no. 4 (1986): 523.

11. Ibid., 527.

12. S. M. Kaczor, "A Tale of Two Gas Stations," *L.U.S.T.Line* (June 2016): 1–4, 24.

13. Ibid., 3.

14. U.S. Environmental Protection Agency, *Underground Storage Tank Program: 25 Years of Protecting Our Land and Water* (Washington, DC: Office of Solid Waste and Emergency Response, 2009), 10.

15. R. Brand, "Taking the Franchising Route to Solve an Environmental Problem," in *True Green: Executive Effectiveness in the U.S. Environmental Protection Agency*, ed. G. A. Emison and J. C. Morris (Lanham: Lexington Books, 2012), 118.

16. "Semiannual Report of UST Performance Measures, End of Fiscal Year 2018," U.S. Environmental Protection Agency, https://www.epa.gov/sites/production/files/2018-11/documents/ca-18-34.pdf.

13. RESURRECTING THE EPA

1. U.S. Environmental Protection Agency, *William D. Ruckelshaus Oral History Interview* (Washington, DC: EPA Oral History Series, 1993), 9.

2. Thomas Jorling, a former Senate Republican staffer who co-authored the Clean Air and Clean Water Acts, as quoted in A. Guillén, "The Radical Idea Behind Trump's EPA Rollbacks," *POLITICO*, June 18, 2017, https://www.politico.com/story/2017/06/18/pruitts-predecessors-pan-epa-originalism-philosophy-239669.

3. Aspen Institute, *EPA 40th Anniversary: 10 Ways EPA Has Strengthened America* (Aspen: Energy and Environment Program, 2010).

4. M. Gunther, "Killing Energy Star: A Popular Program Lands on the Trump Hit List," *Yale Environment 360*, May 4, 2017, https://e360.yale.edu/features/killing-energy-star-a-popular-program-lands-on-the-trump-hit-list.

5. J. A. Mintz, *Enforcement at the EPA* (Austin: University of Texas Press, 1995), 61.

6. J. Goffman, "Reconstruct an Administrative Agency," *The Environmental Forum* 35, no. 6 (2018): 47.

7. C. Giles, "Next Generation Compliance," *The Environmental Forum* 30, no. 5 (2013): 22–26.

8. D. J. Fiorino, *The New Environmental Regulation* (Cambridge: The MIT Press, 2006).

9. "One Water Hub," US Water Alliance, accessed May 2, 2019, http://uswateralliance.org/one-water.

10. P. Baker and C. Davenport, "Obama Orders New Efficiency for Big Trucks," *New York Times*, Feb. 18, 2014, https://www.nytimes.com/2014/02/19/us/politics/obama-to-request-new-rules-for-cutting-truck-pollution.html.

11. R. Leven, "'Do the Opposite Thing You Did 18 Months Ago': EPA Staffers on the Agency in the Trump Era," *Vox*, Nov. 10, 2017, https://www.vox.com/energy-and-environment/2017/11/9/16619988/scott-pruitt-epa-dysfunction-staff.

SELECTED BIBLIOGRAPHY

BOOKS, REPORTS, AND JOURNAL ARTICLES

Alexander, R. B., R. A. Smith, G. E. Schwarz, E. W. Boyer, J. V. Nolan, and J. W. Brakebill. "Differences in Phosphorus and Nitrogen Delivery to the Gulf of Mexico from the Mississippi River Basin." *Environmental Science & Technology* 42, no. 3 (2008): 822–30.

Allaire, M., H. Wu, and U. Lall. "National Trends in Drinking Water Quality Violations." *Proceedings, National Academy of Sciences* 115, no. 9 (2018): 2078–83.

Alley, W. M., and R. Alley. *High and Dry: Meeting the Challenges of the World's Growing Dependence on Groundwater.* New Haven: Yale University Press, 2017.

Allitt, P. A. *Climate of Crisis.* New York: Penguin, 2014.

Alvarez, R. A., D. Zavala-Araiza, D. R. Lyon, D. T. Allen, Z. R. Barkley, A. R. Brandt, K. J. Davis, et al. "Assessment of Methane Emissions from the U.S. Oil and Gas Supply Chains." *Science* 361, no. 6398 (2018): 186–88.

Andrews, D., and B. Walker. *Poisoned Legacy: Ten Years Later, Chemical Safety and Justice for DuPont's Teflon Victims Remain Elusive.* Washington, DC: Environmental Working Group, 2015.

Andrews, R. N. L. *Managing the Environment, Managing Ourselves.* New Haven: Yale University Press, 2006.

Andrews, R. N. L. "The EPA at 40: An Historical Perspective." *Duke Environmental Law & Policy Forum* (Spring 2011): 227–58.

Aspen Institute. *EPA 40th Anniversary: 10 Ways EPA Has Strengthened America.* Aspen: Energy and Environment Program, 2010.

Ayres, G. "A Little Rocket Fuel with Your Salad?" *World Watch Magazine* 16, no. 6 (2003): 12–20.

Batie, S. S. "Wicked Problems and Applied Economics." *American Journal of Agricultural Economics* 90, no. 5 (2008): 1176–91.

Bellar, T. A., J. J. Lichtenberg, and R. C. Kroner. "The Occurrence of Organohalides in Chlorinated Drinking Waters." *Journal of the American Water Works Association* 66, no. 12 (1974): 703–06.

272SELECTED BIBLIOGRAPHY

Bento, A. M., K. Gillingham, M. R. Jacobsen, C. R. Knittel, B. Leard, J. Linn, V. McConnell, et al. "Flawed Analyses of U.S. Auto Fuel Economy Standards." *Science* 362, no. 6419 (2018): 1119–21.

Bolze, D., and J. Beyea. "The Citizen's Acid Rain Monitoring Network." *Environmental Science & Technology* 23, no. 6 (1989): 645–46.

Brand, R. "Taking the Franchising Route to Solve an Environmental Problem." In *True Green: Executive Effectiveness in the U.S. Environmental Protection Agency*, edited by G. A. Emison and J. C. Morris, 111–34. Lanham: Lexington Books, 2012.

Brieger, H. E. "Lust and the Common Law: A Marriage of Necessity." *Boston College Environmental Affairs Law Review* 13, no. 4 (1986): 521–51.

Cannon, J. Z. *Environment in the Balance*. Cambridge: Harvard University Press, 2015.

Centers for Disease Control and Prevention. "Blood Lead Levels in Children Aged 1–5 Years—United States, 1999–2010." *Morbidity and Mortality Weekly Report* 62, no. 13 (2013): 245–48.

Chapelle, F. H. *Ground-Water Microbiology and Geochemistry*. New York: John Wiley, 1993.

Colburn, J. E. "Coercing Collaboration: The Chesapeake Bay Experience." *William & Mary Environmental Law and Policy Review* 40, no. 1 (2016): 678–743.

Copeland, C. *Animal Waste and Water Quality: EPA's Response to the Waterkeeper Alliance Court Decision on Regulation of CAFOs*. Washington DC: Congressional Research Service, 2011.

Cornwall, W. "Critics See Hidden Goal in EPA Data Access Rule." *Science* 360, no. 6388 (2018): 472–73.

Dahl, T. E. *Status and Trends of Wetlands in the Conterminous United States, 2004-2009*. Washington, DC: U.S. Fish and Wildlife Service, 2011.

Desforges, J.-P., A. Hall, B. McConnell, A. Rosing-Asvid, J. L. Barber, A. Brownlow, S. De Guise, et al. "Predicting Global Killer Whale Population Collapse from PCB Pollution." *Science* 361, no. 6409 (2018): 1373–76.

Dieter, C. A., M. A. Maupin, R. R. Caldwell, M. A. Harris, T. I. Ivahnenko, J. K. Lovelace, N. L. Barber, and K. S. Linsey. *Estimated Use of Water in the United States in 2015*. Reston: U.S. Geological Survey, 2018.

Doremus, H. "Scientific and Political Integrity in Environmental Policy." *Texas Law Review* 86 (2008): 1601–53.

Egan, P. J., and M. Mullin. "Climate Change: US Public Opinion." *Annual Review of Political Science* 20 (2017): 209–27.

Erickson, B. E. "Asbestos: Still a Global Menace." *Chemical & Engineering News* 94, no. 47 (2016): 28–31.

Fiorino, D. J. *The New Environmental Regulation*. Cambridge: The MIT Press, 2006.

Freeman, J., and A. Vermeule. "Massachusetts v. EPA: From Politics to Expertise." Harvard Law School Program on Risk Regulation Research Paper No. 08-11 (August 2007). Available at http://ssrn.com/abstract=1307811.

Freemantle, M. *The Chemists' War: 1914–1918*. Cambridge, UK: Royal Society of Chemistry, 2014.

Fretwell, J. D., J. S. Williams, and P. J. Redman, compilers. *National Water Summary on Wetland Resources*. Reston: U.S. Geological Survey, 1996.

Gatz, L. *Waters of the United States (WOTUS): Current Status of the 2015 Clean Water Rule*. Washington, DC: Congressional Research Service, 2018.

Giles, C. "Next Generation Compliance." *The Environmental Forum* 30, no. 5 (2013): 22–26.

Goffman, J. "Reconstruct an Administrative Agency." *The Environmental Forum* 35, no. 6 (2018): 40–47.

Goldman, G. T., and F. Dominici. "Don't Abandon Evidence and Process on Air Pollution Policy." *Science* 363, no. 6434 (2019): 1398–400.

Graham, J. L., N. M. Dubrovsky, and S. M. Eberts. *Cyanobacterial Harmful Algal Blooms and U.S. Geological Survey Science Capabilities*. Reston: U.S. Geological Survey, 2016.

Grandjean, P. "Delayed Discovery, Dissemination, and Decisions on Intervention in Environmental Health: A Case Study on Immunotoxicity of Perfluorinated Alkylate Substances." *Environmental Health* 17, no. 1 (2018): 62–67.

Gurian-Sherman, D. *CAFOs Uncovered: The Untold Costs of Confined Animal Feeding Operations*. Cambridge: Union of Concerned Scientists, 2008.

Hanlon, J., M. Cook, M. Quigley, and B. Wayland. "Water Quality: A Half Century of Progress." EPA Alumni Association, March 25, 2016. Available at http://www.epaalumni.org/hcp/.

Hanna-Attisha, M. *What the Eyes Don't See*. New York: One World, 2018.

Harr, J. *A Civil Action*. New York: Random House, 1996.

Higgins, C. P., and J. A. Field, "Our Stainfree Future? A Virtual Issue on Poly- and Perfluoroalkyl Substances." *Environmental Science & Technology* 51, no. 11 (2017): 5859–60.

Hites, R. A. "Dioxins: An Overview and History." *Environmental Science & Technology* 45, no. 1 (2011): 16–20.

Hogue, C. "Lisa P. Jackson." *Chemical & Engineering News* 88, no. 19 (2010): 14–18.

Hogue, C. "Rocket-Fueled River." *Chemical & Engineering News* 81, no. 33 (2003): 37–46.

Houck, O. A. "Hard Choices: The Analysis of Alternatives Under Section 404 of the Clean Water Act and Similar Environmental Laws." *University of Colorado Law Review* 60 (1989): 773–840.

Hu, X. C., D. Q. Andrews, A. B. Lindstrom, T. A. Bruton, L. A. Schaider, P. Grandjean, R. Lohmann, et al. "Detection of Poly- and Perfluoroalkyl Substances (PFASs) in U.S. Drinking Water Linked to Industrial Sites, Military Fire Training Areas, and Wastewater Treatment Plants." *Environmental Science & Technology Letters* 3, no. 10 (2016): 344–50.

Jacks, G. "Acid Precipitation." In *Regional Ground-Water Quality*, edited by W. M. Alley, 405–21. New York: Wiley, 1993.

Jensen, S. "Report of a New Chemical Hazard." *New Scientist* 32 (1966): 612.

Jones, C. S., J. K. Nielsen, K. E. Schilling, and L. J. Weber. "Iowa Stream Nitrate and the Gulf of Mexico." *PLoS ONE* 13 no. 4 (2018): e0195930.

Kaczor, S. M. "A Tale of Two Gas Stations." *L.U.S.T.Line* (June 2016): 1–4, 24.

Kimm, V. J., J. A. Cotruvo, J. Hoffbuhr, and A. Calvert. "The Safe Drinking Water Act: The First 10 Years." *Journal of the American Water Works Association* 106, no. 8 (2014): 84–95.

Kolok, A. S. *Modern Poisons: A Brief Introduction to Contemporary Toxicology*. Washington, DC: Island Press, 2016.

Koppe, J. G., and J. Keys. "PCBs and the Precautionary Principle." In *The Precautionary Principle in the 20th Century: Late Lessons from Early Warnings*, edited by P. Harremoës, et al., 64–78. New York: Earthscan, 2002.

Kraft, M. E. *Environmental Policy and Politics*. New York: Routledge, 2018.

Landy, M. K., M. J. Roberts, and S. R. Thomas. *The Environmental Protection Agency: Asking the Wrong Questions from Nixon to Clinton*. New York: Oxford University Press, 1994.

Lash, J., D. Sheridan, and K. Gillman. *A Season of Spoils: The Reagan Administration's Attack on the Environment*. New York: Pantheon, 1984.

Lazarus, R. "Who's on First? District, Appeals Courts Grapple with Jurisdiction." *The Environmental Forum* 33, no. 5 (2016): 13.

Lefcheck, J. S., R. J. Orth, W. C. Dennison, D. J. Wilcox, R. R. Murphy, J. Keisman, C. Gurbisz, et al. "Long-Term Nutrient Reductions Lead to the Unprecedented Recovery of a Temperate Coastal Region." *Proceedings of the National Academy of Sciences* 115, no. 14 (2018): 3658–62.

Li, X.-F., and W. A. Mitch. "Drinking Water Disinfection Byproducts (DBPs) and Human Health Effects: Multidisciplinary Challenges and Opportunities." *Environmental Science & Technology* 52, no. 4 (2018): 1681–89.

Likens, G. E., and F. H. Bormann. "Acid Rain: A Serious Regional Environmental Problem." *Science* 184, no. 4142 (1974): 1176–79.

Lim, X. "Tainted Water: The Scientists Tracing Thousands of Fluorinated Chemicals in Our Environment." *Nature* 566, no. 7742 (2019): 26–29.

Macdonald, S. *Propaganda and Information Warfare in the Twenty-First Century: Altered Images and Deception Operations.* New York: Taylor & Francis, 2007.

McCarthy, J. E. *EPA Utility MACT: Will the Lights Go Out?* Washington, DC: Congressional Research Service, 2012.

McCarthy, J. E., and C. Copeland. *EPA Regulations: Too Much, Too Little, or On Track?* Washington, DC: Congressional Research Service, 2016.

McCarthy, J. E., and R. K. Lattanzio. *EPA's 2015 Ozone Air Quality Standards.* Washington, DC: Congressional Research Service, 2017.

McCarthy, J. E., J. L. Ramseur, J. A. Leggett, L. Tsang, and K. C. Shouse. *EPA's Clean Power Plan for Existing Power Plants: Frequently Asked Questions.* Washington, DC: Congressional Research Service, 2017.

McGuire, M. J. *The Chlorine Revolution: Water Disinfection and the Fight to Save Lives.* Denver: American Water Works Association, 2013.

McLellan, E., D. Robertson, K. Schilling, M. Tomer, J. Kostel, and K. King. "Reducing Nitrogen Export from the Corn Belt to the Gulf of Mexico: Agricultural Strategies for Remediating Hypoxia." *Journal of the American Water Resources Association* 51, no. 1 (2015): 263–89

Messner, M., S. Shaw, S. Regli, K. Rotert, V. Blank, and J. Soller. "An Approach for Developing a National Estimate of Waterborne Disease Due to Drinking Water and a National Model Application." *Journal Water Health* 4 (2006): 201–40.

Miller, M. A., R. M. Kudela, A. Mekebri, D. Crane, S. C. Oates, M. T. Tinker, M. Staedler, et al. "Evidence for a Novel Marine Harmful Algal Bloom: Cyanotoxin (Microcystin) Transfer from Land to Sea Otters." *PLoS ONE* 5, no. 9 (2010): e12576.

Mintz, J. A. *Enforcement at the EPA.* Austin: University of Texas Press, 1995.

Nakamura, R. T., and T. W. Church. *Taming Regulation: Superfund and the Challenge of Regulatory Reform.* Washington, DC: Brookings Institution, 2003.

National Academies of Sciences, Engineering, and Medicine. *Valuing Climate Damages: Updating Estimation of the Social Cost of Carbon Dioxide.* Washington, DC: The National Academies Press, 2017.

National Environmental Justice Advisory Council. *Unintended Impacts of Redevelopment and Revitalization Efforts in Five Environmental Justice Communities.* Washington, DC: U.S. Environmental Protection Agency, 2006.

National Research Council. *Alternatives for Managing the Nation's Complex Contaminated Groundwater Sites.* Washington, DC: National Academies Press, 2012.

National Research Council. *Health Implications of Perchlorate Ingestion.* Washington, DC: National Academies Press, 2005.

National Research Council. *In Situ Bioremediation.* Washington, DC: National Academy Press, 1993.

Natural Resources Conservation Service. *Chesapeake Bay Progress Report.* Washington DC: U.S. Department of Agriculture, Sept. 2016.

Office of Management and Budget. *2017 Draft Report to Congress on the Benefits and Costs of Federal Regulations and Agency Compliance with the Unfunded Mandates Reform Act.* Washington, DC: Office of the President of the United States, 2018.

Oreskes, N., and E. M. Conway. *Merchants of Doubt*. New York: Bloomsbury Press, 2010.

Osmond, D. L., D. L. Hoag, A. E. Luloff, D. W. Meals, and K. Neas. "Farmers' Use of Nutrient Management: Lessons from Watershed Case Studies." *Journal of Environmental Quality* 44, no. 2 (2015): 382–90.

Pankow J. F., and J. A. Cherry. *Dense Chlorinated Solvents and other DNAPLs in Groundwater*. Portland: Waterloo Press, 1996.

Perez, M. *Water Quality Targeting Success Stories*. Washington, DC: World Resources Institute and American Farmland Trust, 2017.

Quarles, J. *Cleaning Up America: An Insider's View of the Environmental Protection Agency*. Boston: Houghton Mifflin, 1976.

Revesz, R., and J. Lienke. *Struggling for Air: Power Plants and the "War on Coal."* New York: Oxford University Press, 2016.

Risebrough, R. W., P. Rieche, D. B. Peakall, S. G. Herman, and M. N. Kirven. "Polychlorinated Biphenyls in the Global Ecosystem." *Nature* 220, no. 5172 (1968): 1098–102.

Robin, M.-M. *The World According to Monsanto*. New York: New Press, 2010.

Robinson, M. *Climate Justice*. New York: Bloomsbury, 2018.

Rook, J. J. "The Formation of Halogens during Chlorination of Natural Waters." *Water Treatment and Examination* 23, no. 2 (1974): 234–43.

Ross, B., and S. Amter. *The Polluters*. New York: Oxford University Press, 2010.

Salzman, J. *Drinking Water: A History*. New York: Overlook Duckworth, 2017.

Scavia, D., I Bertani, D. R. Obenour, R. E. Turner, D. R. Forrest, and A. Katin. "Ensemble Modeling Informs Hypoxia Management in the Northern Gulf of Mexico." *Proceedings of the National Academy of Sciences* 114, no. 33 (2017): 8823–28.

Sedlak, D. "Fool Me Once." *Environmental Science & Technology* 50, no. 15 (2016): 7937–38.

Sedlak, D. *Water 4.0: The Past, Present, and Future of the World's Most Vital Resource*. New Haven: Yale University Press, 2014.

Semprini, L., P. K. Kitanidis, D. H. Kampbell, and J. T. Wilson. "Anaerobic Transformation of Chlorinated Aliphatic Hydrocarbons in a Sand Aquifer Based on Spatial Chemical Distributions." *Water Resources Research* 31, no. 4 (1995): 1051–62.

Showstack, R. "Environmental Ratings Lowest Ever for Congressional Republicans." *Eos* 99 (May 2018). Available at https://doi.org/10.1029/2018EO094087.

Steinzor, R., and S. Jones. "Collaborating to Nowhere: The Imperative of Government Accountability for Restoring the Chesapeake Bay." *Journal of Energy & Environmental Law* (Winter 2013): 51–67.

Symons, J. M., T. A. Bellar, J. K. Carswell, J. CeMarco, K. L. Kropp, G. G. Robeck, D. R. Seeger, C. J. Slocum, B. L. Smith, and A. A. Stevens. "National Organics Reconnaissance Survey for Halogenated Organics." *Journal of the American Water Works Association* 67, no. 11 (1975): 634–47.

Tanabe, S. "PCB Problems in the Future: Foresight from Current Knowledge." *Environmental Pollution* 50, no. 1-2 (1988): 5–28.

Tessum, C. W., J. S. Apte, A. L. Goodkind, N. Z. Muller, K. A. Mullins, D. A. Paolella, S. Polasky, et al. "Inequity in Consumption of Goods and Services Adds to Racial–ethnic Disparities in Air Pollution Exposure." *Proceedings of the National Academy of Sciences* 116, no. 13 (2019): 6001–06.

Tiemann, M. *Drinking Water State Revolving Fund (DWSRF): Program Overview and Issues*. Washington, DC: Congressional Research Service, 2017.

Tyer, B. *Opportunity, Montana: Big Copper, Bad Water, and the Burial of an American Landscape*. Boston: Beacon, 2014.

Union of Concerned Scientists. "Fuel Economy and Emissions Standards for Cars and Trucks, Model Years 2017 to 2025." Fact Sheet, June 2016.

U.S. Chemical Safety and Hazard Investigation Board. *Chemical Spill Contaminates Public Water Supply in Charleston, West Virginia.* Washington, DC: Investigation Report 2014-01-I-WV, 2017.

U.S. Environmental Protection Agency. *25 Years of RCRA: Building on Our Past to Protect Our Future.* Washington, DC: Office of Solid Waste and Emergency Response, 2002.

U.S. Environmental Protection Agency. *Addressing Environmental Justice in EPA Brownfields Communities.* Washington, DC: Solid Waste and Emergency Response, 2009.

U.S. Environmental Protection Agency. *Connectivity of Streams and Wetlands to Downstream Waters: A Review and Synthesis of the Scientific Evidence.* Washington, DC: Office of Research and Development, 2015.

U.S. Environmental Protection Agency. *Drinking Water Infrastructure Needs Survey and Assessment, Sixth Report to Congress.* Washington, DC: Office of Water, 2018.

U.S. Environmental Protection Agency. *RCRA's Critical Mission & the Path Forward.* Washington, DC: EPA530-R-14-002, 2014.

U.S. Environmental Protection Agency. *The Benefits and Costs of the Clean Air Act from 1990 to 2020.* Washington, DC: Office of Air and Radiation, 2011.

U.S. Environmental Protection Agency. *The Plain English Guide to the Clean Air Act.* Research Triangle Park: Office of Air Quality Planning and Standards, 2007.

U.S. Environmental Protection Agency. *Superfund Remedy Report 15th Edition.* Washington, DC: 2017.

U.S. Environmental Protection Agency. *Underground Storage Tank Program: 25 Years of Protecting Our Land and Water.* Washington, DC: Office of Solid Waste and Emergency Response, 2009.

U.S. Environmental Protection Agency. *William D. Ruckelshaus Oral History Interview.* Washington, DC: EPA Oral History Series, 1993.

U.S. Government Accountability Office. *Department of Defense Activities Related to Trichloroethylene, Perchlorate, and Other Emerging Contaminants.* Washington, DC: GAO-07-1042T, 2007.

U.S. Government Accountability Office. *EPA Has Increased Efforts to Assess and Control Chemicals but Could Strengthen Its Approach.* Washington, DC: GAO-13-249, 2013.

U.S. Government Accountability Office. *SUPERFUND: Trends in Federal Funding and Cleanup of EPA's Nonfederal National Priorities List Sites.* Washington, DC: GAO-15-812, 2015.

Wiedemeier, T. H., M. A. Swanson, D. E. Moutoux, E. K. Gordon, J. T. Wilson, B. H. Wilson, D. H. Kampbell, et al. *Technical Protocol for Evaluating Natural Attenuation of Chlorinated Solvents in Ground Water.* Washington, DC: U.S. Environmental Protection Agency, 1998.

Wilson, J. T., J. F. McNabb, D. L. Balkwill, and W. C. Ghiorse. "Enumeration and Characterization of Bacteria Indigenous to a Shallow Water-Table Aquifer." *Ground Water* 21, no. 2 (1983): 134–42.

Wilson, J. T., and B. H. Wilson. "Biotransformation of Trichloroethylene in Soil." *Applied and Environmental Microbiology* 49, no. 1 (1985): 242–43.

Zivin, J. G., and M. Neidell. "Air Pollution's Hidden Impacts." *Science* 359, no. 6371 (2018): 39–40.

WEBSITES

Agency for Toxic Substances & Disease Registry. "Toxic Substances Portal—Poly-chlorinated Biphenyls (PCBs)." Accessed April 7, 2019. Available at https://www.atsdr.cdc.gov/ToxProfiles/tp.asp?id=142&tid=26.

Agency for Toxic Substances & Disease Registry. "Toxic Substances Portal—Tri-chloroethylene (TCE)." Accessed April 7, 2019. Available at https://www.atsdr.cdc.gov/phs/phs.asp?id=171&tid=30.

American Rivers. "What Are Prairie Potholes." Accessed April 3, 2019. Available at https://www.americanrivers.org/rivers/discover-your-river/prairie-potholes/.

California Dairy Research Foundation. "California Dairy Quality Assurance Program: Compliance Through Education." Accessed May 2, 2019. Available at http://cdrf.org/home/checkoff-investments/cdqap/.

Chesapeake Bay Foundation. "2016 State of the Bay Report." Accessed May 2, 2019. Available at http://www.cbf.org/about-the-bay/state-of-the-bay-report/2016/index.html.

Climate Leadership Council. "Economists' Statement on Carbon Dividends." Accessed April 21, 2019. Available at https://www.clcouncil.org/economists-statement/.

Environmental Working Group. "Anniston, Alabama: Monsanto Knew About PCB Toxicity for Decades." Accessed April 5, 2019. Available at http://www.ewg.org/research/anniston-alabama/monsanto-knew-about-pcb-toxicity-decades.

U.S. Department of Energy. "Environmental Cleanup." Accessed April 6, 2019. Available at https://energy.gov/national-security-safety/environmental-cleanup.

U.S. Environmental Protection Agency. "8-Hour Ozone (2015) Designated Area/State Information." Accessed April 4, 2019. Available at https://www3.epa.gov/airquality/greenbook/jbtc.html.

U.S. Environmental Protection Agency. "Air Quality - National Summary." Accessed April 4, 2019. Available at https://www.epa.gov/air-trends/air-quality-national-summary.

U.S. Environmental Protection Agency. "Anatomy of Brownfields Redevelopment." Accessed April 6, 2019. Available at https://www.epa.gov/brownfields/anatomy-brownfields-redevelopment

U.S. Environmental Protection Agency. "Brownfields Program Accomplishments and Benefits." Accessed April 6, 2019. Available at https://www.epa.gov/brownfields/brownfields-program-accomplishments-and-benefits.

U.S. Environmental Protection Agency. "Chesapeake Bay TMDL Fact Sheet." Accessed May 2, 2019. Available at https://www.epa.gov/chesapeake-bay-tmdl/chesapeake-bay-tmdl-fact-sheet.

U.S. Environmental Protection Agency. "Deleted National Priorities List (NPL) Sites - by State." Accessed April 7, 2019. Available at https://www.epa.gov/superfund/deleted-national-priorities-list-npl-sites-state.

U.S. Environmental Protection Agency. "EPA Response to Kingston TVA Coal Ash Spill." Accessed April 22, 2019. Available at https://www.epa.gov/tn/epa-response-kingston-tva-coal-ash-spill.

U.S. Environmental Protection Agency. "EPA's PFAS Action Plan." Accessed April 6, 2019. Available at https://www.epa.gov/pfas/epas-pfas-action-plan.

U.S. Environmental Protection Agency. "Hudson River Cleanup." Accessed April 5, 2019. Available at https://www3.epa.gov/hudson/cleanup.html#quest1.

U.S. Environmental Protection Agency. "NPDES CAFO Regulations Implementation Status Reports." Accessed May 2, 2019. Available at https://www.epa.gov/npdes/npdes-cafo-regulations-implementation-status-reports.

U.S. Environmental Protection Agency. "San Gabriel Valley Groundwater Cleanup Superfund Progress Report." May 2017. Available at http://sgvog.org/_assets/2017%20San%20Gabriel%20Valley%20Groundwater%20Cleanup%20Progress%20Report%20FINAL%20%20Updates%205-17-17.pdf.

U.S. Environmental Protection Agency. "Semiannual Report of UST Performance Measures, End of Fiscal Year 2018." Available at https://www.epa.gov/sites/production/files/2018-11/documents/ca-18-34.pdf.

U.S. Environmental Protection Agency. "Superfund Enforcement: 35 Years of Protecting Communities and the Environment." Accessed April 6, 2019. Available at https://www.epa.gov/enforcement/superfund-enforcement-35-years-protecting-communities-and-environment.

U.S. Water Alliance. "One Water Hub." Accessed May 2, 2019. Available at http://uswateralliance.org/one-water.

Water Quality & Health Council. "A Public Health Giant Step: Chlorination of U.S. Drinking Water." May 1, 2008. Available at http://www.waterandhealth.org/drinkingwater/chlorination_history.html.

INDEX

ABOUT THE AUTHORS

Dr. William (Bill) M. Alley is an internationally recognized authority on groundwater and an environmental science writer. He was chief of the Office of Groundwater for the U.S. Geological Survey for almost two decades. Dr. Alley has interacted with the U.S. Environmental Protection Agency in numerous ways for more than forty years, and his experiences allow for an objective, critical look at the agency. **Rosemarie Alley** is a freelance writer with extensive writing and public speaking experience. As a writing team, Bill provides the scientific expertise and Rosemarie makes it interesting and understandable for the general reader. Bill and Rosemarie previously collaborated on *Too Hot to Touch: The Problem of High-Level Nuclear Waste* (2013) and *High and Dry: Meeting the Challenges of the World's Growing Dependence on Groundwater* (2017). The Alleys divide their time between San Diego, California, and Longmont, Colorado.